ACID RAIN:
OVERVIEW AND ABSTRACTS

ACID RAIN:
OVERVIEW AND ABSTRACTS

CARTER N. LANE (EDITOR)

Nova Science Publishers, Inc.
New York

Senior Editors: Susan Boriotti and Donna Dennis
Coordinating Editor: Tatiana Shohov
Office Manager: Annette Hellinger
Graphics: Wanda Serrano and Matt Dallow
Editorial Production: Marius Andronie, Maya Columbus, Vladimir Klestov, Matthew Kozlowski, and Tom Moceri
Circulation: Ave Maria Gonzalez, Vera Popovic, Raymond Davis, Melissa Diaz, Magdalena Nuñez, Marlene Nuñez and Jeannie Pappas
Communications and Acquisitions: Serge P. Shohov
Marketing: Cathy DeGregory

Library of Congress Cataloging-in-Publication Data
Available Upon Request

ISBN 1-59033-461-2.

Copyright © 2003 by Nova Science Publishers, Inc.
400 Oser Ave, Suite 1600
Hauppauge, New York 11788-3619
Tele. 631-231-7269   Fax 631-231-8175
e-mail: Novascience@earthlink.net
Web Site: http://www.novapublishers.com

All rights reserved. No part of this book may be reproduced, stored in a retrieval system or transmitted in any form or by any means: electronic, electrostatic, magnetic, tape, mechanical photocopying, recording or otherwise without permission from the publishers.

The authors and publisher have taken care in preparation of this book, but make no expressed or implied warranty of any kind and assume no responsibility for any errors or omissions. No liability is assumed for incidental or consequential damages in connection with or arising out of information contained in this book.

This publication is designed to provide accurate and authoritative information with regard to the subject matter covered herein. It is sold with the clear understanding that the publisher is not engaged in rendering legal or any other professional services. If legal or any other expert assistance is required, the services of a competent person should be sought. FROM A DECLARATION OF PARTICIPANTS JOINTLY ADOPTED BY A COMMITTEE OF THE AMERICAN BAR ASSOCIATION AND A COMMITTEE OF PUBLISHERS.

Printed in the United States of America

CONTENTS

Preface vii

Title Index ix

Chapter 1 Acid Rain: Background and Overview 1
Environmental Protection Agency

Chapter 2 Implementing Acid Rain Legislation 23
Larry Parker

Chapter 3 Acid Rain, Air Pollution, and Forest Decline 29
Adela Backiel

Bibliography with Abstracts 37

Author Index 155

Subject Index 163

PREFACE

"Acid rain" is a broad term used to describe several ways that acids fall out of the atmosphere. A more precise term is acid deposition, which has two parts: wet and dry. Wet deposition refers to acidic rain, fog, and snow. As this acidic water flows over and through the ground, it affects a variety of plants and animals. The strength of the effects depends on many factors, including how acidic the water is, the chemistry and buffering capacity of the soils involved, and the types of fish, trees, and other living things that rely on the water.

Dry deposition refers to acidic gases and particles. About half of the acidity in the atmosphere falls back to earth through dry deposition. The wind blows these acidic particles and gases onto buildings, cars, homes, and trees. Dry deposited gases and particles can also be washed from trees and other surfaces by rainstorms. When that happens, the runoff water adds those acids to the acid rain, making the combination more acidic than the falling rain alone. Prevailing winds blow the compounds that cause both wet and dry acid deposition across state and national borders, and sometimes over hundreds of miles.

This new book combines an excellent background article with over 900 abstracts and book citations. Easy access is provided by title, author, and subject indexes.

TITLE INDEX

#

^{13}C-NMR spectra and contact time experiment for Skjervatjern fulvic and humic acids, 105

A

A branch exposure chamber for fumigating ponderosa pine to atmospheric pollution, 83

A chemical survey of remote lakes of the Alagnak and Naknek river systems, southwest Alaska, U.S.A., 74

A comparative analysis of aluminum biogeochemistry in a northeastern and a southeastern forested watershed, 58

A comparison of urban and suburban precipitation chemistry, 137

A field study of the oxidation of SO_2 in cloud, 55

A field test of the effect of acidic rain on ion balance in a woodland salamander, 68

A Finnish-Soviet acid rain game: Noncooperative equilibria, cost efficiency, and sulfur agreements, 144

A Mathematical Model to Study the Effect of Toxic Chemicals on aPrey-Predator Type Fishery, 70

A model solution for tracking pollution, 57

A national critical loads framework for atmospheric deposition effects assessment: III. Deposition characterization, 81

A nested grid mesoscale atmospheric chemistry model, 122

A paleolimnological assessment of the effects of post-1970 reductions of sulfur deposition in Sweden, 64

A proposal for the definition of resource equivalency factors for use in product life-cycle assessment, 74

A renewable energy strategy, 138

A study of NO removal by packed-beads discharge reactor, 117

A theoretical investigation of HSO/HOS and their positive ions, 122

Acid air and health, 139

Acid deposition effects on materials: Evaluation of nickel after four years of exposure, 144

Acid depositions and concrete attack: Main influences, 135

Acid fog-induced bronchoconstriction. The role of hydroxymethanesulfonic acid, 40

Acid neutralizing capacity, alkalinity, and acid-base status of natural waters containing organic acids, 79

Acid precipitation and scientific fallout, 113

Acid precursor concentrations above the northeastern United States during summer 1987: Three case studies, 47

Acid rain and the Clean Air Act, 97

Acid rain and the NAPAP study, 37

Acid rain compliance planning: Compliance issues and options, 154

Acid rain compliance: Coordination of state and federal regulation, 117

Acid rain conference held as Congress gives final approval to bill, 152

Acid rain draft study released: No environmental emergency found, 152

Acid rain impacts on utility plans for plant life extension, 44

Acid rain in Asia, 46

Acid rain legislation's complex problem - Fair and efficient emissions limitation, 41

Acid rain mitigation: Everyone benefits from a market for compliance, 62

Acid rain monitoring in Florida from 1978 to the present and evaluation of trends in rainwater composition, 105
Acid rain program offers free-market incentives, portends future regulation, 58
Acid rain reduced in Eastern United States, 103
Acid rain revisited, 37
Acid rain, a world-wide phenomenon: A perspective from the United States, 48
Acid rain, promoting the accumulation of CO_2 in surface waters, 46
Acid rain: What the final bill looks like, 51
Acid rains' dirty business: Stealing minerals from soil, 89
Acid-base characteristics of organic carbon in the HUMEX Lake Skjervatjern, 93
Acidic deposition and global climate change, 116
Acidic deposition, cation mobilization, and biochemical indicators of stress in healthy red spruce, 136
Acidic fog and temperature effects on stigmatic receptivity in two birch species, 85
Acidification and recovery of a spodosol Bs horizon from acidic deposition, 59
Acidification of the HUMEX lake effects on epilimnetic pools and fluxes of carbon, 80
Acidification of the pedosphere, 66
Acidification potential in high mountain lakes in northern New Mexico, 123
Acidification, buffering, and salt effects in the unsaturated zone of a sandy aquifer, Klosterhede, Denmark, 76
Acidity and aluminum toxicity caused by iron oxidation around anode bars, 136
Acidity gradients in the KIPDA region, 70
Action and auction: The SO_2 emissions allowances, 51
Action builds on 1990 Clean Air Act compliance, 127
Addressing the public's goals for environmental regulation when communicating acid rain allowance trades, 102
Adsorption properties of nitrogen monoxide on silver ion-exchanged zeolites, 153
Agenda setting and acid precipitation in the United States, 39
Air pollution and atmospheric precipitation chemistry in Poland - a review, 73
Air pollution damage to U.S. forests, 87
Air quality and land productivity in the northeastern United States, 1980-85, 149
Alaska has 4.0 trillion tons of low-sulfur coal: Is there a future for this resource, 142

Allowance trading made easy: The cash-forward settlement, 42
Allowance trading today and tomorrow: When will it really get started, 111
Allowance trading: Correcting the past and looking to the future, 135
Alternate fuels versus gasoline: A market niche, 72
Aluminum bioconcentration at the gill surface of juvenile Atlantic salmon in acidic media, 150
Aluminum: A neurotoxic product of acid rain, 107
Ambient ozone effects on the ecophysiology of sugar maple (Acer saccharum), 133
An evaluation of the use of dendrochemical analyses in environmental monitoring, 148
An H^+ ion budget approach for evaluating three decades of forest soil acidification, 107
Analysis of corrosion products formed on copper in $Cl_2/H_2S/NO_2$ exposure, 98
Anthropogenic mobilization of sulfur and nitrogen: Immediate and delayed consequences, 69
Application of a LRT model to acid rain control in China, 77
Application of amines for treating flue gas from coal-fired power plants, 96
Application of artificial neural networks in modeling limestone-SO_2 reaction, 41
Aqueous-phase photochemical formation of peroxides in authentic cloud and fog waters, 66
Architecture of the skeletal root system of 40-year-old (Germany), 63
Assessing biogeographic patterns in the changes in soil invertebrate biodiversity due to acidic deposition, 142
Assessment of critical loading of lakes as a basis for remedial measures: A review of fundamental concepts, 76
Associations between forest decline and bird and insect communities in northern hardwoods, 60
Associations between pollutant emissions and precipitation chemistry: An empirical analysis, 54
Atlantic salmon (*Salmo salar*) in winter
Atmospheric chemistry research, 132
Atmospheric corrosion effects of SO_2 and NO_2: A comparison of laboratory and field-exposed copper, 145
Atmospheric corrosion of copper in a rural atmosphere, 117

Atmospheric deposition of nutrients and pollutants in North America: An ecological perspective, 102

Atmospheric dust and acid rain, 79

Attitudes toward environmental hazards: Where do toxic wastes fit, 51

B

Base cation leaching from the canopy of a subtropical rainforest in northeastern Taiwan, 99

Beneficial electrification: Environmental advantages of new electricity uses, 124

Benefit-cost implications of acid rain controls: An evaluation of the NAPAP integrated assessment, 129

Benthic macroinvertebrates along the soil/water interface of the HUMEX lake 1989-1991, 78

Biodesulfurization of coals of different rank: Effect on combustion behavior, 129

Biodiversity and industry ecosystem management, 56

Birch foliar responses to simulated acidic fog and *Septoria betulae* inoculations, 93

Breathing easier: NRDC's work to clear the air, 48

Burner swirls NO_X away, 154

C

CAA update: New HAP, NOx rules: more to come, 52

Calcium declines in northeastern Ontario lakes, 90

CAMRAQ: Comprehensive regional air quality modeling, 77

Can preapproval jump-start the allowance market, 64

Canada, U.S. fight acid rain, 138

Carbon additions increase nitrogen availability in northern hardwood forest soils, 73

Carbon dioxide dynamics in acid forest soils in Shenandoah National Park, Virginia, 53

Care and feeding of continuous emissions monitoring systems, 126

Carryover effects of acid rain and ozone on the physiology of multiple flushes of loblolly pine seedlings, 131

Catching up on the Clean Air Act [news], 109

Cation storage and availability along a Nothofagus forest development sequence in New Zealand, 39

CEMs turn monitoring giant, 149

Cesium stress and adaptation in pseudomonas fluorescens, 40

Changes in poultry litter toxicity with simulated acid rain, 74

Changes in properties of epicuticular wax and the affected by anthropogenic environmental factors, 131

Changes in streamwater chemistry after 20 years from forested watersheds in New Hampshire, U.S.A., 107

Characterization and solubility measurements of uranium-contaminated soils to support risk assessment, 64

Characterization of insoluble fractions of TNT transformed by composting, 53

Chemical rockets and the environment, 45

Chemical trends and status of small lakes near Sudbury, Ontario, 1983-1995: evidence of continued chemical recovery, 106

Chemistry response of two forested watersheds to acid atmospheric deposition, 93

Clean Air Act Amendments and their impacts, 62

Clean air and project financing, 153

Clean air compliance option of choice is low-sulfur coal, 100

Clean air land mine: Continuous monitoring, 149

Clean air law will be costly to chemical industry, 65

Clean coal technologies for gas turbines, 145

Clean coal technology, 37

Clean-burning fuels produced from low-grade coal, 55

Cleaning up coal, 57

Cloud microphysical relationships in California marine stratus, 84

Cloudwater and O_3 effects on red spruce at Whitetop Mt., VA: Physiological response, 121

Coal preparation, 56

Coal preparation: The foundation for modern coal use, 56

Coal: An industry in transition, 108

Cokriging to assess regional stream quality in the Southern Blue Ridge Province, 87

Combating acid deposition and climate change - priorities for Asia, 109

Combined atomic-nuclear-optical-mass techniques for aerosol analysis, 52

Coming out ahead with the Clean Air Act, 43

Common threads: Research lessons from acid rain, ozone depletion, and global warming, 93

Comparison of procedures for preparation of Pinus strobus needle macromolecules, 150

Comparisons between wildfire and forest harvesting and their implications in forest management, 110

Complying with the 1990 Amendments: Managing risks and preparing for prudence reviews, 148

Concentrations of aluminum in gut tissue of crayfish (Procambarus clarkii), purged in sodium chloride, 105

Concentrations of trace elements in recent and preindustrial sediments from Norwegian and Russian Arctic lakes, 137

Congress approves historic clean air legislation, 152

Conservation of exchangeable cations after clear-cutting of a northern hardwood forest, 88

Considerations for establishing relationships between ambient ozone (O_3) and adverse crop response, 94

Considerations for evaluating controlled exposure studies of tree seedlings, 120

Continuous emission monitoring system and its compliance with federal clean air act amendments, 112

Contribution of organic nitrates to the total reactive nitrogen budget at a rural eastern U.S. site, 50

Co-occurrence of ozone and acidic cloudwater in high-elevation forests, 146

Corrosivity mapping -- A novel tool for materials selection and asset management, 92

Court date for EPA acid rain rule, 100

Critical loads and development of acid rain control options, 65

Cross-media approach to saving the Chesapeake Bay, 40

Current, reconstructed past, and projected future status of brook trout (*Salvelinus fontinalis*) streams in Virginia, 51

D

Decision-making for the restoration of Atlantic biology and economics, 123

Design, maintenance extend FGD system slurry valve life, 98

Designation of acid rain and SO_2 control zones and control policies in China, 77

Development of a two-dimensional global tropospheric model: Model chemistry, 83

Dew and frost chemistry at a midcontinent site, United States, 147

Direct and indirect effects of pollution on the foraging behaviour of forest passerines during the breeding season, 49

Distribution of foliar fungal endophytes of *Pinus strobus* between and within host trees, 61

Do acid rain and calcium supply limit eggshell formation for blue tits (Parus caeruleus) in the U.K.?, 125

Does acid rain increase human exposure to mercury A review and analysis of recent literature, 126

Dry deposition of nitrogen dioxide and ozone to coniferous forests, 127

Dry scrubber reduces SO_2 in calciner flue gas, 50

E

Ecological aspects of east-west integration trends, 151

Ecology and the integrated assessment process, 146

Economic incentives for environmental protection: Integrating theory and practice, 75

Economic incentives for optimal sulfur abatement in Europe, 76

Ectomycorrhizal diversity and community structure in oak forest stands exposed to contrasting anthropogenic impacts, 44

EDF: New York oversight plan a disaster for SO_2 trades, 101

Editorial: Acid precipitation, 64

EEI contests N.Y. concerns with allowance trading scheme, 100

Effect of aluminum and zinc on enzyme activities in the green Alga Selenastrum capricorutum, 92

Effect of Clean Air Act Amendments of 1990 on use of Midwestern coal, 61

Effect of pH on leaf decomposition in low alkalinity lakes, 82

Effect of simulated acid rain on the occurrence of Lophodermium on Japanese black pine needles, 41

Effects of acid precipitation on reproduction in alpine plant species, 109

Effects of acid rain and sulfur dioxide on marble dissolution, 134

Effects of acid rain on bird populations, 72

Effects of acidification of metal accumulation by aquatic plants and invertebrates. 2. Wetlands, ponds and small lakes, 38

Effects of acidification on metal accumulation by aquatic plants and invertebrates. 1. Constructed wetlands, 38

Effects of acidity of simulated rain on the fruiting of Summerred' apple trees, 127

Effects of ammonium on elemental nutrition of red spruce and indicator plants grown in acid soil, 82

Effects of environmental mercury on gonadal function in Lake Champlain northern pike (Esox lucius), 68

Effects of excess nitrogen deposition and soil acidification experimental study, 85

Effects of foliar and soil acidity on the rhizospere pH of alfalfa, corn and soybean, 57

Effects of long-term simulated acid rain on suitability of mountain birch for Epirrita autumnata (Geometridae), 143

Effects of ozone and acidic deposition on carbon allocation and mycorrhizal colonization of Pinus taeda L. seedlings, 37

Effects of pretreatment with simulated acid rain on the severity of dogwood anthracnose, 49

Effects of silicate weathering on water chemistry in forested, upland, felsic terrane of the USA, 140

Effects of sorbent injection on particulate properties: Part II. High-temperature sorbent injection, 59

Electrochemical membrane conversion of sodium hydrogen sulfite for the purification of flue gases from sulfur dioxide, 72

Electromotive force responses of Cl_2 gas sensor using $BaCl_2$-KCl solid electrolyte, 40

Electrostatic precipitator electrode upgrade for low sulfur western coal at the OPPD Nebraska City station, 63

Emerald Lake Watershed study: Introduction and site description, 145

Emerging Environmental Issues, 114

Emission control alternatives for electric utility power plants, 54

Emissions dispatch under the underutilization provision of the 1990 U.S. Clean Air Act Amendments: Models and analysis, 81

Emissions of N_2O and NO and net nitrogen nitrogen compounds, 125

Emissions trading -- Market-based approaches offer pollution control incentives, 145

Emissions trading is seen as a new beginning, 44

Emissions trading programs, making sense of the options, 65

Energy and environmental policy: The role of markets, 63

Energy resources law: Update on environmental and health and safety regulatory issues, 92

Energy use and acid deposition: The view from Europe, 141

Environmental aspects of economic relations between nations, 153

Environmental chemistry of small watersheds, 63

Environmental consequences of increased natural-gas usage, 56

Environmental crises of the 21st century: Response and responsibility, 134

Environmental impacts of electricity generation: A global perspective, 124

Environmental trends, 98

Environmental trends: The Tox-man cometh, 98

EPA allocates emission allowances for phase II plants, 51

EPA rates second allowance auction a success, 101

EPA reports results of first SO_2 allowance auction, 65

EPA struggles as 1990 Amendments pass five-year mark, 43

EPA's new emissions trading mechanism: A laboratory evaluation, 53

Equilibrium vapor pressure of H_2O above aqueous H_2SO_4 at low temperature, 108

Erosion by acid rain, accelerating the tracking of polystyrene insulating material, 148

Estimating the corrosion rate of mild steel in sulfuric acid by a hydrogen evolution method, 42

Estimating the flexibility of utility resource plans: An application to natural gas cofiring for SO_2 control, 81

Estimation of regionalized phenomena by geostatistical methods: lake acidity on the Canadian Shield, 44

Evaluation of the direct cost of sulfur abatement under the main desulfurization technologies, 76

Evidence for photochemical formation of H_2O_2 and oxidation of SO_2 in authentic fog water, 87

Experimental study and parameterization of gas absorption by water drops, 39

F

Farm, air, and budget bills await action, 102

Field corn response to acid rain-drought stress interaction, 42
Field measurements of dry deposition compounds using the transition flow reactor, 124
Flue gas desulfurization: the state of the art, 140
Foliar amino acid accumulation as an indicator of ecosystem stress for first-year sugar maple seedlings, 110
Foliar and soil nutrient relationships in red oak and white pine forests, 76
Foliar leaching, translocation, and biogenic emission of ^{35}S in radiolabeled loblolly pines, 69
Foliar nutrient status of Pinus ponderosa exposed to ozone and acid rain, 40
For sale: Sulfur emissions, 79
Force majeure implications of acid rain legislation: The litigation battle of the 1990s, 126
Forest blowdown and lake acidification, 62
Forest Industry: Timberlands tomorrow, 55
Forest soil response to acid and salt additions of sulfate. II. Aluminum and base cations, 60
Formate and acetate in monsoon rainwater of Agra, India, 95
Formation and dissolution kinetics of $Al(OH)_3(s)$ in synthetic freshwater solutions, 103
Formation and evolution of soils from an acidified watershed: Plastic Lake, Ontario, Canada, 92
Fortuitous consequence: The domestic politics of the 1991 Canada-United States agreement on air quality, 39
Fuel cell operation on anaerobic digester gas: Conceptual design and assessment, 140
Fuel type as a factor influencing compliance by utilities under the Acid Rain Programme, 53
Fuels' social costs: Evidence from all electric-generating firms, 115
Fully engineered approach to acid rain control, 43

G

Global atmospheric change and research needs in environmental health sciences, 72
Global outreach, 72
Greater Vancouver Water District drinking water corrosion inhibitor testing, 104
Growth and xylem water potential of white oak and loblolly pine seedlings as affected by simulated acidic rain, 147
Growth response and drought susceptibility of red spruce seedlings exposed to simulated acidic rain and ozone, 98
Growth responses of 53 open-pollinated loblolly pine families to ozone and acid rain, 110

H

H_2O_2 in the marine troposphere and seawater of the Atlantic Ocean (48°N-63°S), 148
Highlights of the Clean Air Act Amendments of 1990, 97
High-surface-area hydrated lime for SO2 control, 128
House clean air markup begins, 152
How brown waters are influenced by acidification: The HUMEX lake case study, 54
How many rooftop PV systems does it take to save the planet?, 91

I

Illinois Power, PacifiCorp protest EPA acid rain plan, 100
Impacts of major watershed perturbations on aquatic ecosystems, 53
Implications of the Clean Air Act acid rain title on industrial boilers, 105
Improve operations and enhance refinery sulfur recovery, 47
Improvement of ecological properties of diesel fuels, 75
In vitro pollen responses of two birch species to acidity and temperature, 85
Increase in the stratospheric background sulfuric acid aerosol mass in the past 10 years, 82
Inexpensive allowances may be costly, 101
Inference of nitrogen cycling in three watersheds of northern Florida, USA, by multivariate statistical analysis, 69
Inferred effects of lake acidification on Daphnia galeata mendotae, 90
Initial step of flue gas desulfurization - an IR study of the reaction of SO_2 with NOx on CaO, 86
Instrumental requirements for global atmospheric chemistry, 38
Intense, natural pollution affects Arctic tundra vegetation at the Smoking Hills, Canada, 68
International environmental law and world order, 75

Investigation of the formation of physical damage on automotive finishes due to acidic reagent exposure, 149

Ion leaching in forest ecosystems along a Great Lakes air pollution gradient, 104

Ionic composition and mineral equilibria of acidic groundwater on the west coast of Sweden, 138

K

Keeping climate research relevant, 129

Keys to fuel supply success, 128

Kinetics and mechanism of the oxidation of HSO_3^- by HSO_5^-, 57

Kinetics and mechanism of the sulfite-induced autoxidation of cobalt(II) in aqueous azide medium, 56

Kinetics of aluminum and sulfate release from forest soil by mono- and diprotic aliphatic acids, 65

Kinetics of chemical weathering in B horizon spodosol fraction, 41

L

Law: Research Journal of the Water Pollution Control Federation, 80

Lichen flora of the Eastern Brook Lakes watershed, Sierra Nevada Mountains, California, 130

Lichen studies along a wet sulfate deposition gradient in Pennsylvania, 137

Limestone treatment of acidified streams, 147

Living in a terrarium, 121

Longitudinal and seasonal water chemistry variations in a northern Appalachian stream, 121

Long-term development of elementa fluxes with bulk precipitation and throughfall in two German forests, 108

Long-term field research on water and environmental quality, 111

Long-term simulation of decreased acid loading on forested watershed, 125

Looking back on SO_2 trading: What's good for the environment is good for the market, 94

Low pH effects on swimming activity of Ambystoma salamander larvae, 95

M

Macroinvertebrate communities in headwater streams affected by acidic precipitation in the central Appalachians, 73

Making a market for SO_2 emissions trading, 139

Managing the global environmental risks in Russia: Missing links and external influences, 138

Marble weathering in an industrial environment, eastern Australia, 63

Materials for photovoltaics, 50

May acid rain legislation excuse performance obligations under coal contracts, 131

Measurement of soil nitrogen oxide emissions at three North American ecosystems, 150

Measurement of toxic and related air pollutants, 87

Measuring OH and HO_2 in the troposphere by laser-induced fluorescence at low pressure, 50

Mechanism of nitrate loss from a forested catchment following a small-scale, natural disturbance, 81

Mechanism of plagioclase dissolution in acid solution at 25°C, 118

Medical responsibility and global environmental change, 108

Mercury and monomethylmercury: Present and future concerns, 67

Method to assess lake responsiveness to future acid inputs using recent synoptic water column chemistry, 152

Microbial extracellular enzyme activities in HUMEX Lake Skjervatjern, 114

Microbial populations in an agronomically managed mollisol treated with simulated acid rain, 111

Mineral element composition of declining and healthy stands of red spruce in western Massachusetts, 82

Mix coal switching with emission trading, 139

Mobile anion concept - Time for a reappraisal, 79

Model simulations of the competing climatic effects of SO_2 and CO_2, 90

Modeling the hydrogeochemical response of a stream to acid deposition using the enhanced trickle-down model, 117

Monsoon shrinks with aerosol models, 113

Mortality of brook trout, mottled sculpins, and slimy sculpins during acidic episodes, 69

Most value planning: Estimating the net benefits of electric utility resource plans, 81

Multi-disciplinary management needs of power systems: A new challenge, 124
Mycorrhizae confer aluminum resistance to tulip-poplar seedlings, 103
Mycorrhization, physiognomy, and first-year survivability Park, 111

N

Nalco technology reduces nitrogen oxide emissions, 128
NAPAP (National Acid Precipitation Assessment Program) results on acid rain, 115
NAPAP: A lesson in science, policy, 130
NARUC winter meetings address key issues for utility industry, 127
NARUC: Regulation at the crossroads, 115
Natural gas as a natural' solution, 109
Natural sulfur emissions to the atmosphere of the continental United States, 43
New form of calcium carbonate improves SO_2 removal from boilers, 116
New method of quantitatively describing drainage areas, 75
New process removes sulfur from coal before burning, 128
New uses for ORNL's ultrasensitive mass spectrometer, 113
New water regulations on the horizon, 112
Nitrate and pH monitoring of rainfall and its effect on west Georgia streams, 88
Nitrate variability in coastal North Carolina rainwater and its impact on the nitrogen cycle in rain, 91
Nitrogen storage and availability during stand development in a New Zealand *Nothofagus* forest, 56
Nitrogen-phosphorus relationship in high mountain lakes: effects of the size of catchment basins, 112
NO release from the isothermal combustion of coal chars, 77
Not rare. But, endangered Elemental profiles of three corticolous lichen species on red spruce in Maine, 142
NOx control techniques for the CPI, 96
NO_X reduction techniques, 86
Nuclear hydrogen - cogeneration and the transitional pathway to sustainable development, 75

O

On a strategy for reducing short-range daytime ground level concentrations due to emissions from very tall stacks, 134
Optimal SO_2 compliance planning using probabilistic production costing and generalized benders decomposition, 84
Optimized acid rain abatement strategies using ecological goals, 43
Opting for tried and true, utilities are overpaying for SO_2 control-NRRI, 101
Organic carbon fractionation applied to lake- and soilwater at the Humex Site, 114
Organic solute changes with acidification in Lake Skjervatjern as shown by ^{1}H-NMR spectroscopy, 106
Over-breeding: Ethically the ultimate environmental problem, 134
Oxidizability and stabilization of ecologically clean diesel fuel, 123
Ozone-induced cytochemical and ultrastructural changes in leaf mesophyll cell walls, 74

P

Performance of Vicia faba plants in relation to simulated acid rain and/or endosulphan treatment, 137
Phase I compliance plans emphasize flexibility, 83
Phase I of acid rain program is a success, 120
Photoacoustic measurement of ammonia in the atmosphere: influence of water vapor and carbon dioxide, 127
Photodissociation of H_2O_2 and CH_3OOH at 248 nm and 298 K: Quantum yields for OH, $O(^3P)$ and $H(^2S)$, 146
Phytoplankton responses to nutrient and grazer manipulations among northeastern lakes of varying pH, 50
Pigouvian taxation of energy for flow and stock externalities and strategic, noncompetitive energy pricing, 151
Planning acid rain compliance: A blend of analytic techniques and engineering judgment, 133
Polluting a microbial methane sink, 122
Pollution control for cash, 92
Pollution solutions in Wonderland, 109
Pollution trading rights could spread, 103
Possible red spruce decline: Contributions of tree-ring analysis, 118

Precipitation chemistry along an inland transect on the Olympic Peninsula, Washington, 46

Precipitation chemistry in and ionic loading to an alpine basin, Sierra Nevada, 150

Precipitation chemistry: Atmospheric loadings to the surface waters of the Indian River lagoon basin by rainfall, 63

Prediction of titratable acidity and soil sensitivity to pH change, 59

Preliminary study of synergism of acid rain and diflubenzuron, 107

Proactive industrial strategies for the Clean Air Act amendments of 1990, 130

Production capital project analysis for the Southern Company, 64

Properties of high fly ash content cellular concrete, 115

Prophecies of doom, 99

Protecting the future: What Congress can do for nuclear power, 151

Pulse energization: A precipitator performance upgrade technology following low sulfur coal switching, 95

Put a lid on NO_x emissions, 59

Putting NO_X in a box, 131

Putting the prices together: The Clean Air Act puzzle, 112

Q

Quantifying the impacts of a national, tradable renewables portfolio standard, 45

Quantitative evaluation of XAD-8 and XAD-4 resins used in tandem for removing organic solutes from water, 106

Question 4: Editorial, 46

R

Rainwater and throughfall chemistry in a terra firme rain forest: Central Amazonia, 67

Red-listed and indicator lichens in woodland key habitats and production forests in Sweden, 88

Reduction of NO with CH_4 effected by copper oxide clusters in the channels of ZSM-5, 84

Regional and microclimatic pollution effects on atmospheric corrosion in Prague and Europe, 92

Regional influences and environmental policymaking: A study of acid rain, 39

Regulatory conflicts facing electric utilities under the Clean Air Act amendments of 1990, 119

Relating sulfate adsorption to soil properties in Michigan forest soils, 104

Release of NO from the combustion of coal chars, 78

Removal of NOx or its conversion into harmless gases by charcoals and composites of metal oxides, 86

Report of the Committee on the Environment, 125, 126

Reproductive response of Great Tits, *Parus major*, in a naturally base-poor forest habitat to calcium supplementation, 106

Reproductive variables of American black1963-1991, 44

Requirements for a viable nuclear future, 128

Response of MICROTOX organisms to leachates of autoclaved cellular concrete, 96

Response of Ned Wilson Lake watershed, Colorado, to changes in atmospheric deposition of sulfate, 52

Response of the Lake Clair Watershed (Duchesnay, Quebec) to changes in precipitation chemistry (1988-1994), 83

Reversing acidification in a forested catchment in southwestern Sweden: Effects on soil solution chemistry, 71

Review and synthesis of experimental data on organic and acidity, 147

S

S_8 threatens natural gas operations, environment, 54

Science or politics NAPAP and Reagan, 102

'Scientific uncertainty' scuttles new acid rain standard, 125

Screening of sorbents for sensors to control sulfur dioxide content in air, 129

Scrubbers are good, 134

Searchers for a new energy source: Tesla, Moray, and Bearden, 89

Second EPA allowance auction set for today, 101

Seed rain and seed bank along an alpine altitudinal gradient in Swedish Lapland, 111

Senate begins clean air legislation debate, 152

Sensitivity of early-life-stage golden trout to low pH and elevated aluminum, 61

Sensitivity of greenback cutthroat trout to acidic pH and elevated aluminum, 151

Sensitivity of twenty soybean cultivars to simulated acid rain, 42

Separating analysis from politics: Acid rain in Europe, 119

Seven summers of energy education courses, 123

Shared resources, common future: Sustainable management of Canada-United States border waters, 130

Shopping for acid rain control strategies, 73

Short range variability of soil chemistry in three acid soils in Ontario, Canada, 68

Short-term responses of wetland vegetation after liming of an Adirondack watershed, 104

Simulated responses of red spruce forest soils to reduced sulfur and nitrogen deposition, 89

SO_2 allowance trading: What rules apply, 138

Soil and soil water studies at the HUMEX site, 146

Soil chemistry and nutrition of North American spruce-fir stands: Evidence of recent change, 89

Soil fauna and site assessment in beech stands of the Belgian Ardennes, 122

Soil solution response to experimentally reduced acid deposition in a forest ecosystem, 38

Soil-water interactions at the HUMEX Lake Skjervatjern, 57

Some aspects of the atmospheric corrosion of copper in the presence of sodium chloride, 142

Some physiological properties of *Cryptomeria japonica* leaves from Kanto, Japan: potential factors causing tree decline, 144

Some reasons for changes in acid deposition during the last decades, 84

Sources of nitrogen in three watersheds of northern Florida, USA: Mainly atmospheric deposition, 69

Spatial variability of canopy throughfall and groundwater sulfate concentrations under a pine stand, 47

Spatial variation in acidic deposition in an appalachian forest, 144

Species extinction mires ecosystem, 82

Spills, drills, and accountability, 140

Spodosol variability and assessment of response to acidic deposition, 60

Stable sulfur isotopes of sulfate in precipitation and stream solutions in a northern hardwood watershed, 140

State inaction could boost clean air costs - EPA, 100

State regulatory issues in acid rain compliance, 139

State Regulatory responses to acid rain: Implications for electric utility operations, 115

Status and future concerns of clinical and environmental aluminum toxicology, 67

Strategy for large scale solubilization of coal-characterization of neurospora protein and gene, 119

Stream chemistry in the eastern United States. 1. Synoptic survey design, acid- base status, and regional patterns, 90

Streams in the New Jersey Pinelands directly reflect changes in atmospheric deposition chemistry, 113

Strength loss in concrete due to varying sulfate exposures, 95

Sulfate corrosion of Portland cement -- Pure and blended with 30% of fly ash, 62

Sulfate deposition over the Arctic Ocean, 138

Sulfate sorption in soils under acid deposition: Comparison of two modeling approaches, 124

Sulfate sorption in soils under acid deposition: Modeling field data from forest liming, 124

Sulfur pool sizes and stable isotope ratios in HUMEX peat before and immediately after the onset of acidification, 113

Sulfur production continues to rise, 118

Sulfuric acid-induced corrosion of aluminum surfaces, 60

Superior Fe-ZSM-5 catalyst for selective catalytic reduction of nitric oxide by ammonia, 102

Survival of brook trout embryos in three episodically acidified streams, 66

Sustainable air quality in the global commons, 78

Swedish scientists take acid-rain research to developing nations, 37

System planning in the 1990s: New choices, new criteria, new solutions, 116

Systematic variations in the concentration of NO_x (NO $+NO_2$) at Niwot Ridge, Colorado, 119

T

Taxing sulfur dioxide emission allowances, 115

Technical efforts focus on cutting LNG plant costs, 40

Ten utilities receive acid rain bonus allowances from EPA, 144

Ten-year study on acid precipitation nears conclusion, 118

Terrestrial mammals of Virginia: Trends in distribution and diversity, 76

The 24 benefits of energy efficiency to electric utilities, 67

The case for a National Association of Physicians for the Environment, 73

The Clean Air Act Amendments of 1990 and industry: Title I non-attainment areas, 132

The Clean Air Act and bonus allowances, 107

The Clean Air Act impacts on rail coal, 136

The collection efficiency of a modified Mohnen slotted-rod cloud-water collector in summer clouds, 91

The consequences of global biomass burning, 99

The cumulative effects of climate warming and other human stresses on Canadian freshwaters in the new millennium, 133

The economics of repowering steam turbines, 93

The effect of H_2O_2 content on the uptake of SO_2 (g) by aqueous droplets, 88

The effects of low pH and elevated aluminum on yellowstone cutthroat trout (Oncorhynchus clarki bouvieri), 66

The effects of recreation disturbance on subalpine seed banks in the Rocky Mountains of Montana, 153

The environmental impact of energy efficiency, 78

The episodic acidification of Adirondack lakes during snowmelt, 132

The erosion of carbonate stone by acid rain: Laboratory and field investigations, 41

The Green Plan: A national challenge for Canada, 118

The greening of urban air, 103

The head dome: A simplified method for human exposures to inhaled air pollutants, 48

The HUMEX Project: Experimental acidification of a catchment and its humic lake, 72

The importance of reporting statistical power: The forest decline and acidic deposition example, 120

The interaction of NO with copper ions in ZSM5: An EPR and IR investigation, 71

The iron catalyzed oxidation of sulfur(IV) in aqueous solution: Differing effects of organics at high and low pH, 107

The long term accuracy and maintainability of an ultrasonic flow monitor on an FGD application, 145

The new Clean Air Act, 118

The outlook for US nuclear power, 111

The question of linkages in environment and development, 114

The rate of sulfite oxidation in seawater, 153

The real cost of energy, 84

The second generation regional acid deposition model chemical mechanism for regional air quality modeling, 141

The solvent extraction approach to petroleum demetallation, 132

The Southern California air quality study, 97

The synergistic effect of hydrogen sulfide and nitrogen dioxide on the atmospheric corrosion of zinc, 143

The use of word-based models to describe the development of UK acid rain policy in the 1980s, 100

The utility industry response to Title IV: generation mix, fuel choice, emissions and costs, 112

Theoretical estimate of the enthalpy of formation of HSO and the HSO-SOH isomerization energy, 151

Thermal and trophic stability of deeper Maine lakes in granite waterhsheds implacted by acid deposition, 141

Three decades of observed soil acidification in the Calhoun Experimental Forest: Has acid rain made a difference?, 106

Total organic carbon in streamwater from four long-term monitored catchments in Norway, 103

Transborder emissions trading between Canada and the United States, 110

Transition metal-catalyzed oxidation of sulfur(IV) oxides. Atmospheric-relevant processes and mechanisms, 48

Transport of octanol soluble carbon and dissolved organic carbon through the soil/water interface of the HUMEX lake, 94

Trends in Pinus ponderosa foliar pigment concentration due to chronic exposure of ozone and acid rain, 116

Tropospheric chemical composition - Overview of experimental methods in measurement, 41

Tropospheric nitrogen: A three-dimensional study of sources, distributions, and deposition, 120

U

U.S. unit to cut sulfur pollution at Polish plant, 135

Uptake and distribution of nitrogen from acidic fog within a ponderosa pine, 66

US acid-rain policy, in court vs. in congress: Heads or tails, 113

US industrial SO_2 emissions expected to remain steady, 146

Use of calibration gases in the US acid rain program, 132

Use of the RAINS model in acid rain negotiations in Europe, 83
Use of TREGRO to simulate the effects of ozone on the growth of red spruce seedlings, 97
User-friendly chemistry takes center stage at ACS meeting, 122
Utilities find it difficult to meet uncertain NO_x control requirements, 141
Utilities swap pollution credits, 148
Utility regulation and the Clean Air Act Amendments of 1990, 101
Utilization of coal combustion by-products: determining the environmental safety, 91

V

Variation in Adirondack, New York, lakewater chemistry as function of surface area, 143
Variation in amount and elemental composition of leaves associated with natural environmental factors, 131
Variation in mineral content of red maple sap across an atmospheric deposition gradient, 108
Variations in the acid-alkali balance of natural waters and some aspects of establishing ecological standards, 86
Vertical leaching of metals from sandy soil minerals in aqueous systems, 70

W

Water flow paths and hydrochemical controls in the Birkenes catchment as inferred from a rainstorm high in seasalts, 114
Weighing environmental externalities: Let's do it right, 89
What happened to science, 77
Where have all the frogs and toads gone, 121
Why aren't there more Atlantic salmon (*Salmo salar*)?, 119
Wind power finding its competitive edge, 89
Wood opportunities, 118
Working with the watershed, 47

Z

Zinc-55% aluminum-1.6% silicon coating compared with zinc coating, 60
Zooplankton community responses to a novel forest insecticide, tebufenozide (RH-5992), in littoral lake enclosures, 94

Chapter 1

ACID RAIN: BACKGROUND AND OVERVIEW[*]

Environmental Protection Agency

Acid rain is a serious environmental problem that affects large parts of the US and Canada. This section of the Web site provides information about acid rain's causes and effects, how we measure acid rain, and what is being done to solve the problem.

WHAT IS ACID RAIN AND WHAT CAUSES IT?

"Acid rain" is a broad term used to describe several ways that acids fall out of the atmosphere. A more precise term is acid deposition, which has two parts: wet and dry.

Wet deposition refers to acidic rain, fog, and snow. As this acidic water flows over and through the ground, it affects a variety of plants and animals. The strength of the effects depend on many factors, including how acidic the water is, the chemistry and buffering capacity of the soils involved, and the types of fish, trees, and other living things that rely on the water.

Dry deposition refers to acidic gases and particles. About half of the acidity in the atmosphere falls back to earth through dry deposition. The wind blows these acidic particles and gases onto buildings, cars, homes, and trees. Dry deposited gases and particles can also be washed from trees and other surfaces by rainstorms. When that happens, the runoff water adds those acids to the acid rain, making the combination more acidic than the falling rain alone.

Prevailing winds blow the compounds that cause both wet and dry acid deposition across state and national borders, and sometimes over hundreds of miles.

Scientists discovered, and have confirmed, that sulfur dioxide (SO_2) and nitrogen oxides (NO_x) are the primary causes of acid rain. In the US, About 2/3 of all SO_2 and 1/4 of all NO_x comes from electric power generation that relies on burning fossil fuels like coal.

---

[*] This section excerpted from the Environmental Protection Agency website: http://www.epa.govairmarkets/acidrain/index.html.

Acid rain occurs when these gases react in the atmosphere with water, oxygen, and other chemicals to form various acidic compounds. Sunlight increases the rate of most of these reactions. The result is a mild solution of sulfuric acid and nitric acid.

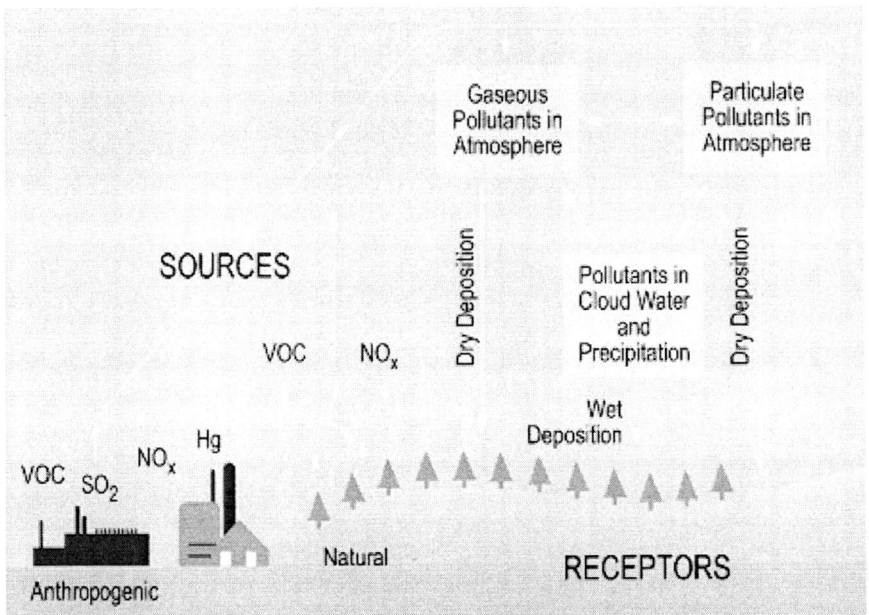

How Do We Measure Acid Rain?

Acid rain is measured using a scale called "pH." The lower a substance's pH, the more acidic it is. See pH for more information.

Pure water has a pH of 7.0. Normal rain is slightly acidic because carbon dioxide dissolves into it, so it has a pH of about 5.5. As of the year 2000, the most acidic rain falling in the US has a pH of about 4.3.

Acid rain's pH, and the chemicals that cause acid rain, are monitored by two networks, both supported by EPA. The National Atmospheric Deposition Program measures wet deposition, and its Web site http://www.epa.gov/epahome/exitepa.htmhttp://www.epa.gov/epahome/exitepa.htmfeatures maps of rainfall pH (follow the link to the isopleth maps) and other important precipitation chemistry measurements.

The Clean Air Status and Trends Network (CASTNET) measures dry deposition. Its Web site features information about the data it collects, the measuring sites, and the kinds of equipment it uses.

What is pH?

Acidic and basic are two extremes that describe chemicals, just like hot and cold are two extremes that describe temperature. Mixing acids and bases can cancel out their extreme effects, much like mixing hot and cold water can even out the water temperature. A substance that is neither acidic nor basic is neutral.

The pH scale measures how acidic or basic a substance is. It ranges from 0 to 14. A pH of 7 is neutral. A pH less than 7 is acidic, and a pH greater than 7 is basic. Each whole pH value below 7 is ten times more acidic than the next higher value. For example, a pH of 4 is ten times more acidic than a pH of 5 and 100 times (10 times 10) more acidic than a pH of 6. The same holds true for pH values above 7, each of which is ten times more alkaline (another way to say basic) than the next lower whole value. For example, a pH of 10 is ten times more alkaline than a pH of 9.

Pure water is neutral, with a pH of 7.0. When chemicals are mixed with water, the mixture can become either acidic or basic. Vinegar and lemon juice are acidic substances, while laundry detergents and ammonia are basic.

Chemicals that are very basic or very acidic are called "reactive." These chemicals can cause severe burns. Automobile battery acid is an acidic chemical that is reactive. Automobile batteries contain a stronger form of some of the same acid that is in acid rain. Household drain cleaners often contain lye, a very alkaline chemical that is reactive.

The following diagram shows the pH scale and the pH of some common items:

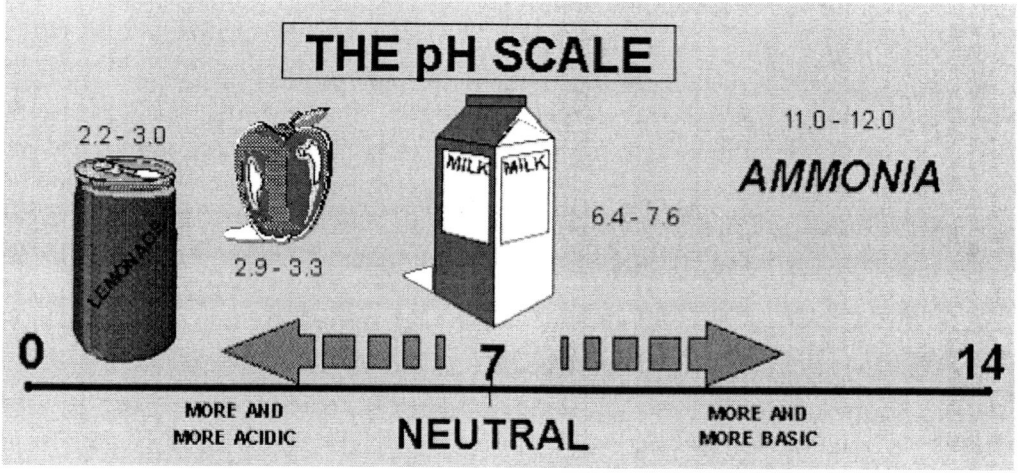

WHAT ARE ACID RAIN'S EFFECTS?

Acid deposition has a variety of effects, including damage to forests and soils, fish and other living things, materials, and human health. Acid rain also reduces how far and how clearly we can see through the air, an effect called visibility reduction. The acid rain effects section provides more details on each of these.

Effects of Acid Rain

Acid rain causes acidification of lakes and streams and contributes to damage of trees at high elevations (for example, red spruce trees above 2,000 feet) and many sensitive forest soils. In addition, acid rain accelerates the decay of building materials and paints, including irreplaceable buildings, statues, and sculptures that are part of our nation's cultural heritage.

Prior to falling to the earth, SO_2 and NO_x gases and their particulate matter derivatives, sulfates and nitrates, contribute to visibility degradation and harm public health.

Effects of Acid Rain: Lakes & Streams

The ecological effects of acid rain are most clearly seen in the aquatic, or water, environments, such as streams, lakes, and marshes. Acid rain flows to streams, lakes, and marshes after falling on forests, fields, buildings, and roads. Acid rain also falls directly on aquatic habitats. Most lakes and streams have a pH between 6 and 8, although some lakes are naturally acidic even without the effects of acid rain. Acid rain primarily affects sensitive bodies of water, which are located in watersheds whose soils have a limited ability to neutralize acidic compounds (called "buffering capacity"). Lakes and streams become acidic (pH value goes down) when the water itself and its surrounding soil cannot buffer the acid rain enough to neutralize it. In areas where buffering capacity is low, acid rain also releases aluminum from soils into lakes and streams; aluminum is highly toxic to many species of aquatic organisms.

Where Does Acid Rain Affect Lakes and Streams?

Many lakes and streams examined in a National Surface Water Survey (NSWS) suffer from chronic acidity, a condition in which water has a constant low pH level. The survey investigated the effects of acidic deposition in over 1,000 lakes larger than 10 acres and in thousands of miles of streams believed to be sensitive to acidification. Of the lakes and streams surveyed, acid rain caused acidity in 75 percent of the acidic lakes and about 50 percent of the acidic streams. Several regions in the U.S. were identified as containing many of the surface waters sensitive to acidification. They include the Adirondacks and Catskill Mountains in New York state, the mid-Appalachian highlands along the east coast, the upper Midwest, and mountainous areas of the Western United States. In areas like the Northeastern United States, where soil buffering capacity is poor, some lakes now have a pH value of less than 5. One of the most acidic lakes reported is Little Echo Pond in Franklin, New York. Little Echo Pond has a pH of 4.2.

Acidification is also a problem in lakes that were not surveyed in federal research projects. For example, although lakes smaller than 10 acres were not included in the NSWS, there are from one to four times as many of these small lakes as there are larger lakes. In the Adirondacks, the percentage of acidic lakes is significantly higher when it includes smaller lakes.

Streams flowing over soil with low buffering capacity are as susceptible to damage from acid rain as lakes. Approximately 580 of the streams in the Mid-Atlantic Coastal Plain are acidic primarily due to acidic deposition. In the New Jersey Pine Barrens, for example, over 90 percent of the streams are acidic, which is the highest rate of acidic streams in the nation. Over 1,350 of the streams in the Mid-Atlantic Highlands (mid-Appalachia) are acidic, primarily due to acidic deposition.

The acidification problem in both the United States and Canada grows in magnitude if "episodic acidification" is taken into account. Episodic acidification refers to brief periods during which pH levels decrease due to runoff from melting snow or heavy downpours. Lakes and streams in many areas throughout the United States are sensitive to episodic acidification. In the Mid-Appalachians, the Mid-Atlantic Coastal Plain, and the Adirondack Mountains, many additional lakes and streams become temporarily acidic during storms and spring snowmelt. For example, approximately 70 percent of sensitive lakes in the Adirondacks are at risk of episodic acidification. This amount is over three times the amount of chronically acidic lakes. In the mid-Appalachians, approximately 30 percent of sensitive streams are likely to become acidic during an episode. This level is seven times the number of chronically acidic streams in that area. Episodic acidification can cause "fish kills."

Emissions from U.S. sources also contribute to acidic deposition in eastern Canada, where the soil is very similar to the soil of the Adirondack Mountains, and the lakes are consequently extremely vulnerable to chronic acidification problems. The Canadian government has estimated that 14,000 lakes in eastern Canada are acidic.

How Does Acid Rain Affect Fish and Other Aquatic Organisms?

Acid rain causes a cascade of effects that harm or kill individual fish, reduce fish population numbers, completely eliminate fish species from a waterbody, and decrease biodiversity. As acid rain flows through soils in a watershed, aluminum is released from soils into the lakes and streams located in that watershed. So, as pH in a lake or stream decreases, aluminum levels increase. Both low pH and increased aluminum levels are directly toxic to fish. In addition, low pH and increased aluminum levels cause chronic stress that may not kill individual fish, but leads to lower body weight and smaller size and makes fish less able to compete for food and habitat.

Some types of plants and animals are able to tolerate acidic waters. Others, however, are acid-sensitive and will be lost as the pH declines. Generally, the young of most species are more sensitive to environmental conditions than adults. At pH 5, most fish eggs cannot hatch. At lower pH levels, some adult fish die. Some acid lakes have no fish. The chart below shows that not all fish, shellfish, or the insects that they eat can tolerate the same amount of acid; for example, frogs can tolerate water that is more acidic (has lower pH) than trout.

Trout	pH 5.0
Bass	pH 5.5
Perch	pH 4.5
Frogs	pH 4.0
Salamanders	pH 5.0
Clams	pH 6.0
Crayfish	pH 5.5
Snails	pH 6.0
Mayfly	pH 5.5

How Does Acid Rain Affect Ecosystems?

Together, biological organisms and the environment in which they live are called an ecosystem. The plants and animals living within an ecosystem are highly interdependent. For example, frogs may tolerate relatively high levels of acidity, but if they eat insects like the mayfly, they may be affected because part of their food supply may disappear. Because of the connections between the many fish, plants, and other organisms living in an aquatic ecosystem, changes in pH or aluminum levels affect biodiversity as well. Thus, as lakes and streams become more acidic, the numbers and types of fish and other aquatic plants and animals that live in these waters decrease.

What is the Role of Nitrogen in Acid Rain and other Environmental Problems?

The impact of nitrogen on surface waters is also critical. Nitrogen plays a significant role in episodic acidification and new research recognizes the importance of nitrogen in long-term chronic acidification as well. Furthermore, the adverse impact of atmospheric nitrogen deposition on estuaries and near-coastal water bodies is significant. Scientists estimate that from 10-45 percent of the nitrogen produced by various human activities that reaches estuaries and coastal ecosystems is transported and deposited via the atmosphere. For example, about 30 percent of the nitrogen in the Chesapeake Bay comes from atmospheric deposition. Nitrogen is an important factor in causing eutrophication (oxygen depletion) of water bodies. The symptoms of eutrophication include blooms of algae (both toxic and non-toxic), declines in the health of fish and shellfish, loss of seagrass beds and coral reefs, and ecological changes in food webs. According to the National Oceanic and Atmospheric Administration, these conditions are common in many of our nation's coastal ecosystems. These ecological changes impact human populations by changing the availability of seafood and creating a risk of consuming contaminated fish or shellfish, reducing our ability to use and enjoy our coastal ecosystems, and causing economic impact on people who rely on healthy coastal ecosystems, such as fishermen and those who cater to tourists.

How is the Acid Rain Program Addressing these Issues?

Acid rain control will produce significant benefits in terms of lowered surface water acidity. If acidic deposition levels were to remain constant over the next 50 years (the time frame used for projection models), the acidification rate of lakes in the Adirondack Mountains that are larger than 10 acres would rise by 50 percent or more. Scientists predict, however, that the decrease in SO_2 emissions required by the Acid Rain Program will significantly reduce acidification due to atmospheric sulfur. Without the reductions in SO_2 emissions, the proportions of acidic aquatic ecosystems would remain high or dramatically worsen.

Effects of Acid Rain: Forests

Over the years, scientists, foresters, and others have watched some forests grow more slowly without knowing why. The trees in these forests do not grow as quickly at a healthy pace. Leaves and needles turn brown and fall off when they should be green and healthy. In extreme cases, individual trees or entire areas of the forest simply die off without an obvious reason.

Researchers now know that acid rain causes slower growth, injury, or death of forests. Acid rain has been implicated in forest and soil degradation in many areas of the eastern United States, particularly high elevation forests of the Appalachian Mountains from Maine to Georgia that include areas such as the Shenandoah and Great Smoky Mountain National Parks. Of course, acid rain is not the only cause of such conditions. Other things that add stress, such as air pollutants, insects, disease, drought, or very cold weather also harm trees and plants. In most cases, in fact, the impacts of acid rain on trees occur due to the combined effects of acid rain and these other environmental stressors. After many years of collecting information on the chemistry and biology of forests, researchers are beginning to understand how acid rain works on the forest soil, trees, and other plants.

Acid Rain on the Forest Floor

A spring shower in the forest washes leaves and falls through the trees to the forest floor below. Some trickles over the ground and runs into a stream, river, or lake, and some of the water soaks into the soil. That soil may neutralize some or all of the acidity of the acid rainwater. This ability is called buffering capacity, and without it, soils become more acidic. Differences in soil buffering capacity are an important reason why some areas that receive acid rain show a lot of damage, while other areas that receive about the same amount of acid rain do not appear to be harmed at all. The ability of forest soils to resist, or buffer, acidity depends on the thickness and composition of the soil, as well as the type of bedrock beneath the forest floor. Midwestern states like Nebraska and Indiana have soils that are well buffered. Places in the mountainous northeast, like New York's Adirondack and Catskill Mountains, have thin soils with low buffering capacity.

How Acid Rain Harms Trees

Acid rain does not usually kill trees directly. Instead, it is more likely to weaken trees by damaging their leaves, limiting the nutrients available to them, or exposing them to toxic substances slowly released from the soil. Quite often, injury or death of trees is a result of these effects of acid rain in combination with one or more additional threats.

Scientists know that acidic water dissolves the nutrients and helpful minerals in the soil and then washes them away before trees and other plants can use them to grow. At the same time, acid rain causes the release of substances that are toxic to trees and plants, such as aluminum, into the soil. Scientists believe that this combination of loss of soil nutrients and increase of toxic aluminum may be one way that acid rain harms trees. Such substances also wash away in the runoff and are carried into streams, rivers, and lakes. More of these substances are released from the soil when the rainfall is more acidic.

However, trees can be damaged by acid rain even if the soil is well buffered. Forests in high mountain regions often are exposed to greater amounts of acid than other forests because they tend to be surrounded by acidic clouds and fog that are more acidic than rainfall. Scientists believe that when leaves are frequently bathed in this acid fog, essential nutrients in their leaves and needles are stripped away. This loss of nutrients in their foliage makes trees more susceptible to damage by other environmental factors, particularly cold winter weather.

How Acid Rain Affects Other Plants

Acid rain can harm other plants in the same way it harms trees. Although damaged by other air pollutants such as ground level ozone, food crops are not usually seriously affected because farmers frequently add fertilizers to the soil to replace nutrients that have washed away. They may also add crushed limestone to the soil. Limestone is an alkaline material and increases the ability of the soil to act as a buffer against acidity.

The Effects of Acid Rain on Automotive Coatings

Over the past two decades, there have been numerous reports of damage to automotive paints and other coatings. The reported damage typically occurs on horizontal surfaces and appears as irregularly shaped, permanently etched areas. The damage can best be detected under fluorescent lamps, can be most easily observed on dark colored vehicles, and appears to occur after evaporation of a moisture droplet. In addition, some evidence suggests damage occurs most frequently on freshly painted vehicles. Usually the damage is permanent; once it has occurred, the only solution is to repaint.

The general consensus within the auto industry is that the damage is caused by some form of environmental fallout. "Environmental fallout," a term widely used in the auto and coatings industries, refers to damage caused by air pollution (e.g., acid rain), decaying insects, bird droppings, pollen, and tree sap. The results of laboratory experiments and at least one field study have demonstrated that acid rain can scar automotive coatings. Furthermore, chemical analyses of the damaged areas of some exposed test panels showed elevated levels of sulfate, implicating acid rain.

The popular term "acid rain" refers to both wet and dry deposition of acidic pollutants that may damage material surfaces, including auto finishes. These pollutants, which are released when coal and other fossil fuels are burned, react with water vapor and oxidants in the atmosphere and are chemically transformed into sulfuric and nitric acids. The acidic compounds then may fall to earth as rain, snow, fog, or may join dry particles and fall as dry deposition. Automotive coatings may be damaged by all forms of acid rain, including dry deposition, especially when dry acidic deposition is mixed with dew or rain. However, it has been difficult to quantify the specific contribution of acid rain to paint finish damage relative to damage caused by other forms of environmental fallout, by the improper application of paint or by deficient paint formulations. According to coating experts, trained specialists can differentiate between the various forms of damage, but the best way of determining the cause of chemically induced damage is to conduct a detailed, chemical analysis of the damaged area.

Because evaporation of acidic moisture appears to be a key element in the damage, any steps taken to eliminate its occurrence on freshly painted vehicles may alleviate the problem. The steps include frequent washing followed by hand drying, covering the vehicle during precipitation events, and use of one of the protective coatings currently on the market that claim to protect the original finish. (However, data on the performance of these coatings are not yet sufficient.)

The auto and coatings industries are fully aware of the potential damage and are actively pursuing the development of coatings that are more resistant to environmental fallout, including acid rain. The problem is not a universal one-- it does not affect all coatings or all vehicles even in geographic areas known to be subject to acid rain-- which suggests that technology exists to protect against this damage. Until that technology is implemented to protect all vehicles or until acid deposition is adequately reduced, frequent washing and drying and covering the vehicle appear to be the best methods for consumers who wish to minimize acid rain damage.

Effects of Acid Rain: Materials

Acid rain and the dry deposition of acidic particles contribute to the corrosion of metals (such as bronze) and the deterioration of paint and stone (such as marble and limestone). These effects seriously reduce the value to society of buildings, bridges, cultural objects (such as statues, monuments, and tombstones), and cars.

Dry deposition of acidic compounds can also dirty buildings and other structures, leading to increased maintenance costs. To reduce damage to automotive paint caused by acid rain and acidic dry deposition, some manufacturers use acid-resistant paints, at an average cost of $5 for each new vehicle (or a total of $61 million per year for all new cars and trucks sold in the U.S.) The Acid Rain Program will reduce damage to materials by limiting SO_2 emissions. The benefits of the Acid Rain Program are measured, in part, by the costs now paid to repair or prevent damage--the costs of repairing buildings and bridges, using acid-resistant paints on new vehicles, plus the value that society places on the details of a statue lost forever to acid rain.

To observe the effects of acid rain on marble and limestone, two building materials commonly used in monuments, ancient buildings, and in many modern structures:

- Place a piece of chalk in a bowl with white vinegar.
- Place another piece in a bowl of tap water.
- Leave the dishes overnight.

The next day, see if you can tell which piece of chalk is more worn away.

This experiment with chalk allows you to see the effect of acid rain on marble and limestone because chalk is made of calcium carbonate, a compound occurring in rocks, such as marble and limestone, and in animal bones, shells, and teeth.

Effects of Acid Rain: Visibility Reduction

Sulfates and nitrates that form in the atmosphere from sulfur dioxide (SO_2) and nitrogen oxides (NO_x) emissions contribute to visibility impairment, meaning we can't see as far or as clearly through the air. Sulfate particles account for 50 to 70 percent of the visibility reduction in the eastern part of the United States, affecting our enjoyment of national parks, such as the Shenandoah and the Great Smoky Mountains. The Acid Rain Program is expected to improve the visual range in the eastern U.S. by 30 percent. Based on a study of the value national park visitors place on visibility, the visual range improvements expected at national parks of the eastern United States due to the Acid Rain Program's SO_2 reductions will be worth over a billion dollars annually by the year 2010. In the western part of the United States, nitrates and carbon also play roles, but sulfates have been implicated as an important source of visibility impairment in many of the Colorado River Plateau national parks, including the Grand Canyon, Canyonlands, and Bryce Canyon.

Effects of Acid Rain: Human Health

Acid rain looks, feels, and tastes just like clean rain. The harm to people from acid rain is not direct. Walking in acid rain, or even swimming in an acid lake, is no more dangerous than walking or swimming in clean water. However, the pollutants that cause acid rain (sulfur dioxide (SO_2) and nitrogen oxides (NO_x)) also damage human health. These gases interact in the atmosphere to form fine sulfate and nitrate particles that can be transported long distances by winds and inhaled deep into people's lungs. Fine particles can also penetrate indoors. Many scientific studies have identified a relationship between elevated levels of fine particles and increased illness and premature death from heart and lung disorders, such as asthma and bronchitis.

Based on health concerns, SO_2 and NO_x have historically been regulated under the Clean Air Act, including the Acid Rain Program. In the eastern United States, sulfate aerosols make up about 25 percent of fine particles. By lowering SO_2 and NO_x emissions from power generation, the Acid Rain Program will reduce the levels of fine sulfate and nitrate particles and so reduce the incidence and the severity of these health problems. When fully implemented by the year 2010, the public health benefits of the Acid Rain Program are estimated to be valued at $50 billion annually, due to decreased mortality, hospital admissions, and emergency room visits.

Decreases in nitrogen oxide emissions are also expected to have a beneficial impact on human health by reducing the nitrogen oxides available to react with volatile organic compounds and form ozone. Ozone impacts on human health include a number of morbidity and mortality risks associated with lung inflammation, including asthma and emphysema.

How Do We Reduce Acid Rain?

What Society Can Do About Acid Deposition

There are several ways to reduce acid deposition, more properly called acid deposition, ranging from societal changes to individual action.

Understand Acid Deposition's Causes and Effects

To understand acid deposition's causes and effects and track changes in the environment, scientists from EPA, state governments, and academic study acidification processes. They collect air and water samples and measure them for various characteristics like pH and chemical composition, and they research the effects of acid deposition on human-made materials such as marble and bronze. Finally, scientists work to understand the effects of sulfur dioxide (SO_2) and nitrogen oxides (NO_x) - the pollutants that cause acid deposition and fine particles - on human health.

To solve the acid rain problem, people need to understand how acid rain causes damage to the environment. They also need to understand what changes could be made to the air pollution sources that cause the problem. The answers to these questions help leaders make better decisions about how to control air pollution and therefore how to reduce - or even eliminate - acid rain. Since there are many solutions to the acid rain problem, leaders have a choice of which options or combination of options are best. The next section describes some of the steps that can be taken to reduce, or even eliminate, the acid deposition problem.

Clean up Smokestacks and Exhaust Pipes

Almost all of the electricity that powers modern life comes from burning fossil fuels like coal, natural gas, and oil. acid deposition is caused by two pollutants that are released into the atmosphere, or emitted, when these fuels are burned: sulfur dioxide (SO_2) and nitrogen oxides (NO_x).

Coal accounts for most US sulfur dioxide (SO_2) emissions and a large portion of NO_x emissions. Sulfur is present in coal as an impurity, and it reacts with air when the coal is burned to form SO_2. In contrast, NO_x is formed when any fossil fuel is burned.

There are several options for reducing SO_2 emissions, including using coal containing less sulfur, washing the coal, and using devices called scrubbers to chemically remove the SO_2 from the gases leaving the smokestack. Power plants can also switch fuels; for example burning natural gas creates much less SO_2 than burning coal. Certain approaches will also have additional benefits of reducing other pollutants such as mercury and carbon dioxide.

Understanding these "co-benefits" has become important in seeking cost-effective air pollution reduction strategies. Finally, power plants can use technologies that don't burn fossil fuels. Each of these options has its own costs and benefits, however; there is no single universal solution.

Similar to scrubbers on power plants, catalytic converters reduce NO_x emissions from cars. These devices have been required for over twenty years in the US, and it is important to keep them working properly and tailpipe restrictions have been tightened recently. EPA has also made, and continues to make, changes to gasoline that allows it to burn cleaner.

Use Alternative Energy Sources

There are other sources of electricity besides fossil fuels. They include: nuclear power, hydropower, wind energy, geothermal energy, and solar energy. Of these, nuclear and hydropower are used most widely; wind, solar, and geothermal energy have not yet been harnessed on a large scale in this country.

There are also alternative energies available to power automobiles, including natural gas powered vehicles, battery-powered cars, fuel cells, and combinations of alternative and gasoline powered vehicles.

All sources of energy have environmental costs as well as benefits. Some types of energy are more expensive to produce than others, which means that not all Americans can afford all types of energy. Nuclear power, hydropower, and coal are the cheapest forms today, but changes in technologies and environmental regulations may shift that in the future. All of these factors must be weighed when deciding which energy source to use today and which to invest in for tomorrow.

Restore a Damaged Environment

Acid deposition penetrates deeply into the fabric of an ecosystem, changing the chemistry of the soil as well as the chemistry of the streams and narrowing, sometimes to nothing, the space where certain plants and animals can survive. Because there are so many changes, it takes many years for ecosystems to recover from acid deposition, even after emissions are reduced and the rain becomes normal again. For example, while the visibility might improve within days, and small or episodic chemical changes in streams improve within months, chronically acidified lakes, streams, forests, and soils can take years to decades or even centuries (in the case of soils) to heal.

However, there are some things that people do to bring back lakes and streams more quickly. Limestone or lime (a naturally-occurring basic compound) can be added to acidic lakes to "cancel out" the acidity. This process, called liming, has been used extensively in Norway and Sweden but is not used very often in the United States. Liming tends to be expensive, has to be done repeatedly to keep the water from returning to its acidic condition, and is considered a short-term remedy in only specific areas rather than an effort to reduce or prevent pollution. Furthermore, it does not solve the broader problems of changes in soil chemistry and forest health in the watershed, and does nothing to address visibility reductions, materials damage, and risk to human health. However, liming does often permit fish to remain in a lake, so it allows the native population to survive in place until emissions reductions reduce the amount of acid deposition in the area.

Look to the Future

As emissions from the largest known sources of acid deposition - power plants and automobiles-are reduced, EPA scientists and their colleagues must assess the reductions to make sure they are achieving the results Congress anticipated. If these assessments show that acid deposition is still harming the environment, Congress may begin to consider additional ways to reduce emissions that cause acid deposition. They may consider additional emissions reductions from sources that have already been controlled, or methods to reduce emissions from other sources. They may also invest in energy efficiency and alternative energy. The cutting edge of protecting the environment from acid deposition will continue to develop and implement cost-effective mechanisms to cut emissions and reduce their impact on the environment.

Take Action as Individuals

It may seem like there is not much that one individual can do to stop acid deposition. However, like many environmental problems, acid deposition is caused by the cumulative actions of millions of individual people. Therefore, each individual can also reduce their contribution to the problem and become part of the solution. One of the first steps is to understand the problem and its solutions.

Individuals can contribute directly by conserving energy, since energy production causes the largest portion of the acid deposition problem. For example, you can:

- Turn off lights, computers, and other appliances when you're not using them
- Use energy efficient appliances: lighting, air conditioners, heaters, refrigerators, washing machines, etc.
- Only use electric appliances when you need them.
- Keep your thermostat at 68 F in the winter and 72 F in the summer. You can turn it even lower in the winter and higher in the summer when you are away from home.
- Insulate your home as best you can.
- Carpool, use public transportation, or better yet, walk or bicycle whenever possible
- Buy vehicles with low NO_x emissions, and maintain all vehicles well.
- Be well-informed.

ACID RAIN PROGRAM: OVERVIEW

The overall goal of the Acid Rain Program is to achieve significant environmental and public health benefits through reductions in emissions of sulfur dioxide (SO_2) and nitrogen oxides (NO_x), the primary causes of acid rain. To achieve this goal at the lowest cost to society, the program employs both traditional and innovative, market-based approaches for controlling air pollution. In addition, the program encourages energy efficiency and pollution prevention.

Phases and Reductions

Title IV of the Clean Air Act set a goal of reducing annual SO_2 emissions by 10 million tons below 1980 levels. To achieve these reductions, the law required a two-phase tightening of the restrictions placed on fossil fuel-fired power plants.

Phase I began in 1995 and affected 263 units at 110 mostly coal-burning electric utility plants located in 21 eastern and midwestern states. An additional 182 units joined Phase I of the program as substitution or compensating units, bringing the total of Phase I affected units to 445. Emissions data indicate that 1995 SO_2 emissions at these units nationwide were reduced by almost 40% below their required level.

Phase II, which began in the year 2000, tightened the annual emissions limits imposed on these large, higher emitting plants and also set restrictions on smaller, cleaner plants fired by coal, oil, and gas, encompassing over 2,000 units in all. The program affects existing utility units serving generators with an output capacity of greater than 25 megawatts and all new utility units.

The Act also called for a 2 million ton reduction in NO_x emissions by the year 2000. A significant portion of this reduction has been achieved by coal-fired utility boilers that will be required to install low NO_x burner technologies and to meet new emissions standards.

Operating Principles: Feasible, Flexible, Accountable

The Acid Rain Program is implemented through an integrated set of rules and guidance designed to accomplish three primary objectives:

1. Achieve environmental benefits through reductions in SO_2 and NO_x emissions.
2. Facilitate active trading of allowances and use of other compliance options to minimize compliance costs, maximize economic efficiency, and permit strong economic growth.
3. Promote pollution prevention and energy efficient strategies and technologies.

Each individual component fulfills a vital function in the larger program:

- the allowance trading system creates low-cost rules of exchange that minimize government intrusion and make allowance trading a viable compliance strategy for reducing SO_2
- the opt-in program allows nonaffected industrial and small utility units to participate in allowance trading
- the NO_x emissions reduction rule sets new NO_x emissions standards for existing coal-fired utility boilers and allows emissions averaging to reduce costs
- the permitting process affords sources maximum flexibility in selecting the most cost-effective approach to reducing emissions
- the continuous emission monitoring (CEM) requirements provide credible accounting of emissions to ensure the integrity of the market-based allowance system and to verify the achievement of the reduction goals

- the excess emissions provision provides incentives to ensure self-enforcement, greatly reducing the need for government intervention
- the appeals procedures allow the regulated community to appeal decisions with which it may disagree

Together these measures ensure the achievement of environmental benefits at the least cost to society.

Environmental Benefits

Acid rain causes acidification of lakes and streams and contributes to damage to trees and many sensitive forest soils. In addition, acid rain accelerates the decay of building materials and paints, including irreplaceable buildings, statues, and sculptures that are part of our nation's cultural heritage. Prior to falling to the earth, SO_2 and NO_x gases and their particulate matter derivatives, sulfates and nitrates, contribute to visibility degradation and impact public health.

The Acid Rain Program confers significant benefits on the nation. By reducing SO_2 and NO_x, many acidified lakes and streams will significantly improve so that they can once again support fish life. Visibility will improve, allowing for increased enjoyment of scenic vistas across our country, particularly in National Parks. Stress to our forests that populate the ridges of mountains from Maine to Georgia will be reduced. Deterioration of our historic buildings and monuments will be slowed. Most importantly, reductions in SO_2 and NO_x will reduce fine particulate matter (sulfates, nitrates) and ground level ozone (smog), leading to improvements in public health.

Allowance Trading

The Acid Rain Program represents a dramatic departure from traditional command and control regulatory methods which establish specific, inflexible emissions limitations with which all affected sources must comply. Instead, the Acid Rain Program introduces an allowance trading system that harnesses the incentives of the free market to reduce pollution.

Under this system, affected utility units are allocated allowances based on their historic fuel consumption and a specific emissions rate. Each allowance permits a unit to emit 1 ton of SO_2 during or after a specified year. For each ton of SO_2 emitted in a given year, one allowance is retired, that is, it can no longer be used.

Allowances may be bought, sold, or banked. Anyone may acquire allowances and participate in the trading system. However, regardless of the number of allowances a source holds, it may not emit at levels that would violate federal or state limits set under Title I of the Clean Air Act to protect public health.

During Phase II of the program (now in effect), the Act set a permanent ceiling (or cap) of 8.95 million allowances for total annual allowance allocations to utilities. This cap firmly restricts emissions and ensures that environmental benefits will be achieved and maintained.

Annual Reconciliation

Annual reconciliation is the process by which EPA compares a regulated unit's annual emissions and the number of allowances it owns. At the end of each year, units are granted a 60-day grace period to ensure that they have sufficient allowances to match their SO_2 emissions during the previous year. If they need to, they may buy allowances during the grace period. Units may sell allowances that exceed their emissions or bank them for use in future years.

The Allowance Tracking System

EPA has instituted an electronic recordkeeping and notification system called the Allowance Tracking System (ATS) to track allowance transactions and the status of allowance accounts. ATS is the official tally of allowances by which EPA determines compliance with the emissions limitations. Any party interested in participating in the trading system may open an ATS account by submitting an application to EPA. Accounts contain information on unit account balances, account representatives (which must be appointed by each trading party), and serial numbers for each allowance. ATS is computerized to expedite the flow of data and to assist in the development of a viable market for allowances.

Auctions and Direct Sale

EPA holds an allowance auction annually. The auctions help to send the market an allowance price signal, as well as furnish utilities with an additional avenue for purchasing needed allowances. The direct sale offered allowances at a fixed price of $1,500 (adjusted for inflation). Anyone could buy allowances in the direct sale, but independent power producers (IPPs) could obtain written guarantees from EPA stating that they had first priority. These guarantees, which were awarded on a first-come, first-served basis, secured the option for qualified IPPs to purchase a yearly amount of allowances over a 30 year span. This provision enabled IPPs to assure lenders that they would have access to the allowances they needed to build and operate new units. The direct sale was eliminated in 1997 because this provision proved to be unnecessary.

Voluntary Entry: The Opt-in Program

The Opt-in Program expands EPA's Acid Rain Program to include additional sulfur dioxide (SO_2) emitting sources. Recognizing that there are additional emission reduction opportunities in the industrial sector, Congress established the Opt-in Program under section 410 of the Clean Air Act Amendments of 1990. The Opt-in Program allows sources not required to participate in the Acid Rain Program the opportunity to enter the program on a voluntary basis and receive their own SO_2 allowances.

The participation of these additional sources will reduce the cost of achieving the 10 million ton reduction in SO_2 emissions mandated under the Clean Air Act. As participating sources reduce their SO_2 emissions at a relatively low cost, their reductions -- in the form of allowances -- can be transferred to electric utilities where emission reductions are more expensive.

The Opt-in Program offers a combustion source a financial incentive to voluntarily reduce its SO_2 emissions. By reducing emissions below its allowance allocation, an opt-in source will have unused allowances, which it can sell in the SO_2 allowance market. Opting in will be profitable if the revenue from the sale of allowances exceeds the combined cost of the emissions reduction and the cost of participating in the Opt-in Program.

Pollution Prevention

The allowance trading system contains an inherent incentive for utilities to prevent pollution, since for each ton of SO_2 that a utility avoids emitting, one fewer allowance must be retired. Utilities that reduce emissions through energy efficiency and renewable energy are able to sell, use, or bank their surplus allowances. As also provided in the Act, EPA has set aside a reserve of 300,000 allowances to stimulate energy efficiency and renewable energy generation. Those utilities that either implement demand-side energy conservation programs to curtail emissions or install renewable energy generation facilities may be eligible to receive bonus allowances from this reserve.

Nitrogen Oxides (NO_x) Reductions

The Clean Air Act Amendments of 1990 set a goal of reducing NO_x by 2 million tons from 1980 levels. The Acid Rain program focuses on one set of sources that emit NO_x, coal-fired electric utility boilers. As with the SO_2 emission reduction requirements, the NO_x program was implemented in two phases, beginning in 1996 and 2000.

The NO_x program embodies many of the same principles of the SO_2 trading program, in that it also has a results-oriented approach, flexibility in the method to achieve emission reductions, and program integrity through measurement of the emissions. However, it does not "cap" NO_x emissions as the SO_2 program does, nor does it utilize an allowance trading system.

Emission limitations for the NO_x boilers provide flexibility for utilities by focusing on the emission rate to be achieved (expressed in pounds of NO_x per million Btu of heat input). In general, two options for compliance with the emission limitations are provided:

- compliance with an individual emission rate for a boiler
- averaging of emission rates over two or more units to meet an overall emission rate limitation

These options give utilities flexibility to meet the emission limitations in the most cost-effective way and allow for the further development of technologies to reduce the cost of compliance.

If a utility properly installs and maintains the appropriate control equipment designed to meet the emission limitation established in the regulations, but is still unable to meet the limitation, the NO_x program allows the utility to apply for an alternative emission limitation (AEL) that corresponds to the level that the utility demonstrates is achievable.

Phase I of the NO_x program began on January 1, 1996 and applied to two types of boilers (which were already targeted for Phase I SO_2 reductions): dry-bottom wall-fired boilers and tangentially fired boilers. Dry-bottom wall-fired boilers had to meet a limitation of 0.50 lbs of NO_x per mmBtu averaged over the year, and tangentially fired boilers had to achieve a limitation of 0.45 lbs of NO_x per mmBtu, again, averaged over the year. Approximately 170 boilers needed to comply with these NO_x performance standards during Phase I.

Phase II of the NO_x program began in 2000. These regulations:

1. set lower emission limits for Group 1 boilers first subject to an acid rain emissions limitation in Phase II, and
2. established initial NO_x emission limitations for Group 2 boilers, which include boilers applying cell-burner technology, cyclone boilers, wet bottom boilers, and other types of coal-fired boilers.

The final rule was promulgated December 19, 1996.

Emissions Monitoring and Reporting

Under the Acid Rain Program, each unit must continuously measure and record its emissions of SO_2, NO_x, and CO_2, as well as volumetric flow and opacity. In most cases, a continuous emission monitoring (CEM) system must be used. There are provisions for initial equipment certification procedures, periodic quality assurance and quality control procedures, recordkeeping and reporting, and procedures for filling in missing data periods. Units report hourly emissions data to EPA on a quarterly basis. This data is then recorded in the Emissions Tracking System, which serves as a repository of emissions data for the utility industry. The emissions monitoring and reporting systems are critical to the program. They instill confidence in allowance transactions by certifying the existence and quantity of the commodity being traded and assure that NO_x averaging plans are working. Monitoring also ensures, through accurate accounting, that the SO_2 and NO_x emissions reduction goals are met.

Excess Emissions

If annual emissions exceed the number of allowances held, the owners or operators of delinquent units must pay a penalty of $2,000 (adjusted for inflation) per excess ton of SO_2 or NO_x emissions. In addition, violating utilities must offset the excess SO_2 emissions with allowances in an amount equivalent to the excess. A utility may either have allowances deducted immediately or submit an excess emissions offset plan to EPA that outlines how these cutbacks will be achieved.

Designated Representatives

Each source appoints one individual, the Designated Representative, to represent the owners and operators of the source in all matters relating to the holding and disposal of allowances for its units that are affected by the Clean Air Act. The Designated Representative is also responsible for all submissions pertaining to permits, compliance plans, emission monitoring reports, offset plans, compliance certification, and other necessary information. A source may appoint an Alternate Designated Representative to act on behalf of the Designated Representative.

Permitting

The Designated Representative for each source is required to file an acid rain permit application for the source and a compliance plan to the Title V permitting authority for each affected unit at the source. The Acid Rain permits and compliance plans are simple, allow sources to fashion a compliance strategy tailored to their individual needs, and foster trading. For example, they allow sources to make real-time allowance trading decisions through the use of automatic permit amendments.

Acid rain permits, which are also issued by the relevant Title V permitting authority, require that each unit account hold a sufficient number of allowances to cover the unit's SO_2 emissions in each year, comply with the applicable NO_x limit, and monitor and report emissions. Permits are subject to public comment before approval.

Compliance Options: Freedom to Choose

The Acid Rain Program allows sources to select their own compliance strategy. For example, to reduce SO_2 an affected source may repower its units, use cleaner burning fuel, or reassign some of its energy production capacity from dirtier units to cleaner ones. Sources also may decide to reduce electricity generation by adopting conservation or efficiency measures. Most options, like fuel switching, require no special prior approval, allowing the source to respond quickly to market conditions without needing government approval. For NO_x, the source may meet the performance standard on a utility-unit basis, enter into an emissions averaging plan, or apply for an alternative emissions limitation.

In either case, the program allows affected utilities to combine these and other options in ways they see fit in order to tailor their compliance plans to the unique needs of each unit or system.

A Model Program

EPA gained broad input into the development of the Acid Rain Program by consulting with representatives from various stakeholder groups, including utilities, coal and gas companies, emissions control equipment vendors, labor, academia, Public Utility Commissions, state pollution control agencies, and environmental groups.

EPA is maintaining this open door policy as it implements the program, and it continues to solicit ideas from the numerous and diverse individuals and groups interested in acid rain control. In addition, EPA is collaborating with groups who wish to evaluate the benefits and effects of the program through economic and environmental studies.

The Acid Rain Program is already being viewed around the world as a prototype for tackling emerging environmental issues. The allowance trading system capitalizes on the power of the marketplace to reduce SO_2 emissions in the most cost-effective manner possible. The permitting program allows sources the flexibility to tailor and update their compliance strategy based on their individual circumstances. The continuous emissions monitoring and reporting systems provide the accurate accounting of emissions necessary to make the program work, and the excess emissions penalties provide strong incentives for self-enforcement. Each of these separate components contributes to the effective working of an integrated program that lets market incentives do the work to achieve cost-effective emissions reductions. The General Accounting Office recently confirmed the beneftis of this approach, projecting that the allowance trading system could save as much as $3 billion per year -- over 50% -- compared with a command and control approach typical of previous environmental protection programs.

GLOSSARY

A

Acid Deposition
The process by which acidic particles, gases, and precipitation leave the atmosphere. More commonly referred to as acid rain, acid deposition has two components: wet and dry deposition.

acid rain
The result of sulfur dioxide (SO_2) and nitrogen oxides (NO_x) reacting in the atmosphere with water and returning to earth as rain, fog, or snow. Broadly used to include both wet and dry deposition. The acid rain page provides a great deal of information about this issue.

Al
Aluminum; a metal that is toxic to trees and fish

allowance
A tradeable permit to emit a specific amount of a pollutant. For example, under the Acid Rain Program, one allowance permits the emissions of one ton of sulfur dioxide (SO2).

anions
Negatively charged molecule such as sulfate ($SO4(2-)$) and nitrate ($NO3-$). In combination with hydrogen ($H+$), these molecules act as strong acids.

acid neutralizing capacity (ANC)

A measure of the ability for water or soil to neutralize added acids. This is done by the reaction of hydrogen ions with inorganic or organic bases such as bicarbonate (HCO3-) or organic ions.

acidification
Refers to reducing something's pH, making it more acidic; also means the loss of ANC.

adsorb
To take up and hold (a gas, liquid, or dissolved substance) in a thin layer of molecules on the surface of a solid substance.

B

buffering capacity
The resistance of water or soil to changes in pH.

base cations
Positively charged ions such as magnesium, sodium, potassium, and calcium that increase pH of water (make it less acidic) when released to solution through mineral weathering and exchange reactions.

C

Ca(2+)
Calcium; a base cation that helps to reduce acidification

chronic acidification
Generally refers to surface waters that remain acidified (ANC<0) regardless of variations in hydrologic conditions (precipitation, stream flow, etc.).

D

deposition
The processes by which chemical constituents move from the atmosphere to the earth's surface. These processes include precipitation (wet deposition, such as rain or cloud fog), as well as particle and gas deposition (dry deposition).

dose response functions
The relationship between the effects (response) on an organism or system and the amount (dose) of some material to which the organism/system is exposed.

dry deposition
The settling of gases and particles out of the atmosphere. Dry deposition is a component of acid deposition, more commonly referred to as acid rain.

E

eutrophication
A reduction in the amount of oxygen dissolved in water. The symptoms of eutrophication include blooms of algae (both toxic and non-toxic), declines in the health of fish and shellfish, loss of seagrass beds and coral reefs, and ecological changes in food webs.

L

leaching
Process by which water removes chemicals from soil through chemical reactions and the downward movement of water.

M

Mg(2+)
Magnesium; a base cation that helps to reduce acidification.

mineral weathering
The physical and chemical breakdown of rocks that releases ions such as calcium and aluminum.

MW
Megawatt; a unit for describing how much electricity a power plant can generate. The Acid Rain Program includes virtually all units in the US that can generate over 25 MW.

N

nitrogen fixation
The process in which bacteria convert biologically unusable nitrogen gas (N2) into biologically usable ammonia (NH3) and nitrates (NO3-).

nitrogen oxides (NO_x)
A group of gases that cause acid rain and other environmental problems, such as smog and eutrophication of coastal waters. Burning fossil fuels, such as coal and gasoline, releases NO_x into the atmosphere. Various programs are reducing NO_x emissions, including the Acid Rain Program and NO_x cap and trade programs.

P

pH
A scale that denotes how acidic or basic a substance is. Pure water has a pH of 7.0 and is neither acidic nor basic.

precipitation

Water in the form of rain, sleet, or snow (wet deposition).

S

sulfur dioxide (SO_2)

A gas that causes acid rain. Burning fossil fuels, such as coal, releases SO_2 into the atmosphere. Various EPA programs are reducing SO_2 emissions, including the Acid Rain Program.

W

wet deposition

The process by which chemicals are removed from the atmosphere and deposited on the Earth's surface via rain, sleet, snow, cloudwater, and fog.

Chapter 2

IMPLEMENTING ACID RAIN LEGISLATION

Larry Parker

INTRODUCTION

The broad-ranging provisions in Title IV of The Clean Air Act Amendments of 1990 (P.L. 101-549) raise myriad implementation issues, particularly with respect to the system of tradable "allowances." This system provides an economic mechanism by which emitters of sulfur dioxide (SO_2) can determine the most cost-effective way to meet reduction requirements. Issues include the role of States, particularly public utilities commissions; unfinished rulemaking, including defining repowering technologies and developing revised fossil fuel utility SO_2 and NO_x New Source Performance Standards (NSPS); and operational issues, including pricing of allowances, auctions, and the Phase 1 allowance pool.

The allowance system represents a new turn in environmental policy in directly addressing the issue of pollution control costs. In doing so, it presumes that cost-effectiveness will be the driving factor in SO_2 control decisionmaking by utilities, State regulators, and Federal agencies. While the economics of this new system has been extensively debated, the legal, regulatory, and industry structure contexts in which the new system of tradable allowances will operate may need further definition if various conflicts develop. For example, the industry structure is designed to focus first on the reliability of electricity supply; to the extent that the uncertainty of the allowance system interferes with that priority, it may be bypassed.

Title IV, Acid Deposition Control, sets goals for Jan. 1, 2000, of reducing annual sulfur dioxide (SO_2) emissions by 10 million tons and annual nitrogen oxides (NO_x) emissions by 2.0 million tons from 1980 levels. Beginning in the year 2000, total utility 502 emissions are limited to 8.9 million tons and total industrial SO_2 emissions are limited to 5.6 million tons.

Title IV is implemented through a two-phase process. Under Phase 1, owners l operators of 110 facilities listed in the law are required to reduce emissions by about 3.5 million tons annually by Jan. 1, 1995. Under Phase 2, all utility facilities greater than 25 megawatts (Mw) are affected by the bill. To maintain emission reductions beyond the Phase 2 compliance date

(Jan. 1, 2000), the law essentially caps SO_2 emissions at existing individual sources through a tonnage limitation, and at future plants through the offset requirement.

The mechanism for achieving the reductions is a comprehensive permit and emission allowance system. An allowance is a limited authorization to emit a ton of SO_2. Facilities receive allowances based on specific formulas contained in the law. These allowances may be traded or banked for future use or sale. Allowance sales and auctions are to be held to ensure liquidity in the allowance market.

Despite debate about the economics of this new system and the addition in the statute of various provisions -- such as the auction systems-- to improve the potential for success of the allowance system, difficulties and conflicts have emerged. However, current indications suggest that complying with Title IV will be considerably less expensive than the costs estimated by the utility industry or government during the legislative debate on these provisions.

DEVELOPMENTS

In March 1995, the Chicago Board of Trade held its third annual allowance auction. A total of 176400 allowances were sold with prices averaging between $128 and $132 an allowance. This represents a drop from the $150 an allowance average price received from the 1994 auction, and confirms a downward trend with respect to Title IV compliance costs.

In March1995, the EPA announced its final rule on allowing industrial boilers (but not including industrial processes) to "opt-in" to the allowance program. The program is strictly voluntary and facilities choosing to enter must comply with monitoring, enforcement, and other CAA requirements similar to those for utility participants.

In February 1995, the EPA released a draft report on the feasibility of a acid deposition standard. Required by the CAA Amendments of 1990, the draft study suggests that the mandated reductions under Title IV will provide clear benefits to sensitive surface water, that nitrogen dioxide, as well as sulfur dioxide, plays an important role in acidification of surface waters, and that the allowance trading program is not projected to harm surface waters.

BACKGROUND AND ANALYSIS

P.L. 101-549, Title IV -- Provisions of Acid Rain Law

The broad-ranging provisions in Title IV of The Clean Air Act Amendments of 1990 (P.L. 101-549) raise myriad implementation issues. As enacted, Title Iv sets goals for Jan. 1, 2000, of an annual reduction of 10 million tons of SO_2 and of 2.0 million tons of NO_x emissions below 1980 levels. Total utility SO_2 emissions are limited to 8.9 million tons beginning in 2000 and total industrial SO_2 emissions are limited to 5.6 million tons. (Total SO_2 emissions in 1980 were approximately 24 million tons; utility SO_2 emissions were approximately 16 million tons.)

Implementation -- SO₂

Title IV is being implemented through a two-phase process. For Phase 1, the statute lists 110 of the largest and most polluting electric utility facilities; those have a generating capacity of more than 100 megawatts (Mw) and emit at a rate greater than 2.5 lb. of SO_2 per million Btus (mmbtu). By Jan. 1, 1995, the owners/operators of these facilities must curtail SO_2 emissions based on a 2.5 lb. SO_2 per mmbtu rate times their annual average fuel consumption during 1985-87 (called the baseline). It is anticipated that this will reduce SO_2 emission by about 3.5 million tons. (See Table 1.)

Phase 2 lowers the threshold emissions rate from 2.5 lb. to 1.2 lb. per mmBtu and the facility size affected to 75 Mw, which will increase the number of affected facilities by about an order of magnitude (these facilities are often called the "big-dirties," or Group 1 units). In addition, powerplants at lower emissions rates and smaller facilities (between 25 and 75 Mw), also face limits on emissions.

Generally, powerplants emitting at less than 1.2 lb. of 502 per mmbtu are permitted to multiply the lesser of the actual or allowable 1985 SO_2 emission rate by their 1985-87 baseline times 1.2 (then divided by 2000 to convert into tons), or replace their 1985-87 baseline and the 1.2 multiplier with a baseline number based on fuel consumption at a 60% capacity factor (from 2000-2009 only; after that, the first calculation is the only option).

Small units (from 25 Mw to 75 Mw) must meet tonnage limitations based on either 1.2 lb. of SO_2 per mmBtu times their 1985-87 baseline if the utility system is greater than 250 Mw or the lesser of their actual or allowable 1985 SO_2 rate times their 1985-87 baseline if the utility system is less than 250 Mw.

Units under 25 Mw and simple combustion turbines are exempted from the reduction requirements, although they may opt in the program if they so desire.

Overall, when the Phase 2 reductions are in place on Jan. 1, 2000, total emissions should meet the statutory-required limits.

Table 1. SO₂ Utility Emission Reductions Under P.L. 101-549

	Facility (emission rate = lbs. per mmbtu)	Comply By	Reduction/Cap
Phase 1	>100Mw and >2.5 lb. SO₂ (2.5 x baseline)*	Jan. 1, 1995	3.5 million ton reduction
Phase 2	>75Mw and >1.2 lb. SO₂ (1.2 x baseline)*	Jan. 1, 2000	8.9 million ton cap
	>75Mw and <1.2 lb. (lesser of actual or allowable emission rate in 1985 x baseline x 1.2 or use different baseline based on consumption at a 60% factor -- latter option from 2000 to 2009 only)*		
	25Mw to 75 Mw (either 1.2 x baseline if whole system is >250Mw or lesser of actual or allowable 1985 rate x base-line if whole system is <250Mw)*		
	(<25Mw exempt from reductions)		
After Phase 2	all facilities >25Mw		cap existing units; offsets for new units; NSPS for new units

* Baseline = annual average fuel consumption during 1985-1987. Formulas require division by 2000 to convert to tons. Note that there are many special cases where different formulas, specified in the law, apply.

Allowances

To introduce some flexibility in the distribution and timing of reductions, the law creates a comprehensive permit and emissions allowance system. An allowance is a limited authorization to emit a ton of SO_2. Each affected facility must have as many allowances as it has tons of SO_2 emissions for a calendar year; facilities receive allowances either under statutory formulas, or by purchase or trade in a market to be set up by EPA regulation.

For existing powerplants -- those in operation at enactment -- allowances are issued by EPA; the allowances will be allocated in accordance with the formulas discussed above (Table 1). New powerplants -- those which commence operation after enactment -- will receive no allowances, except under specific limited circumstances. (Also, as discussed later, EPA is to revise New Source Performance Standards (NSPS) for future powerplants.) In order to operate after the year 2000, when emissions are capped, these new units will have to obtain allowances (offsets) from those units allocated allowances. The allowances may be traded nationally during either phase, or may be banked for later use or sale. The law also permits industrial sources and powerplants to go under the allowance system and to sell allowances to utility systems under regulations to be developed by EPA.

NO_x-for- SO_2 trading is to be studied, but is not permitted.

Special Adjustments to Allowances

Powerplants with emissions rates between 1.2 lb. and 2.5 lb. may receive an adjustment to their allowance allocation if they have low 1985-87 baseline numbers. In addition, units that commenced or commence operation between Jan. 1, 1986, and Dec. 31, 1995, have special formulas or allocations for their allowances.

Because the acid rain program particularly affects the Midwest (which has high SO_2 emissions and will bear high cleanup costs) and the West (which has low 802 emissions and hence few allowances that could be made available to account for growth), the Act redistributes a total of approximately 8.9 million SO_2 allowances to the Midwest in Phase 1 and to clean units in Phase 2. Specifically, about 3.5 million of these allowances are earmarked for Phase 1 powerplants choosing to install 90% control technology. Such units may delay Phase 1 compliance from 1995 to 1997 and receive two allowances for each ton of SO_2 reduced below a 1.2 lb. per mmBtu level during 1997-1999. In addition, during Phase 1, 200,000 allowances annually are to be distributed to units (except Department of Energy-contracted facilities) in Illinois, Indiana, and Ohio. The specific sources of these allowances are not specified in the law, and are not included in the 3.5 million ton reserve identified above.

For Phase 2, about 5.3 million additional allowances are allocated over a 10-year period to relatively clean plants (2000-2009). These provisions generally allow most facilities below 1.2 lbs. of SO_2 per mmbtu to increase use by about 20%, allocate some additional allowances to low-medium SO_2 emitters with low baseline calculations, and grant some allowances for predominately natural gas-fired units. In addition, the Act provides a special option for States that have State-wide emission rates of 0.8 lbs. per mmBtu or less. Finally, 50,000 permanent allowances are allocated annually among 10 Midwestern States. These 50,000 additional allowances are not subject the 8.9 million ton emissions cap.

Sales/Auctions

Beginning in 1993, the law provides for sales and auctions to improve the liquidity of the allowance system and ensure the availability of allowances for utilities and independent power producers who need them. A fund consisting of 2.8% of Phase 1 and Phase 2 allowance allocations is set aside for sale. From 1993 to 1999, 25,000 allowances are to be sold annually at a fixed price of $1,500 an allowance; beginning in 2000, the sale will offer 50,000 allowances annually. Independent power producers have guaranteed rights to these allowances under certain conditions. In addition, 150,000 allowances will be offered annually at auction during 1993-1995, and 250,000 from 1996-1999. This will be an open auction with no minimum price. Utilities with excess allowances may have them auctioned off, and any person may buy allowances.

Emissions Caps

To maintain emission reductions beyond the Phase 2 compliance dates, the law essentially caps 802 emissions at individual existing sources through a tonnage limitation, and at future plants through the allowance system. First, emissions from most existing sources are capped at a specified emission rate times an historic average fuel consumption level. Second, since plants commencing operation after enactment must obtain allowances for any 802 emissions, with minor exceptions their emissions must effectively be offset by additional reductions at existing facilities allocated allowances. The utility 802 emission cap is set at 8.9 million tons, with some exceptions, just over half the 1988 utility emissions of approximately 15 million tons.

The law also requires EPA to inventory industrial emissions of 802 and to report every 5 years, beginning in 1995. If the inventory shows that industrial emissions may reach levels above 5.60 million tons per year, then EPA is to take action under the Act to ensure that the 5.60 million ton cap is not exceeded (1988 industrial emissions of 802 were approximately 5.9 million tons).

Implementation -- NO_x

The law requires EPA to set specific NO_x emission rate limitations --0.45 lb. per mmBtu for tangentially fired boilers and 0.50 lb. per mmBtu for wall-fired boilers--unless those rates can not be achieved by low-NO_x burner technology. Tangentially and wall-fired boilers affected by Phase 1 SO_2 controls must also meet NO_x requirements. EPA is to set emission limitations for other types of boilers by 1997 based on low- NO_x burner costs. In addition, EPA is to propose and promulgate a revised New Source Performance Standard (NSPS) for NO_x by Jan. 1, 1994.

Excess Emissions

The law provides that if an affected unit does not have sufficient allowances to cover its emissions, it is subject to an excess emission penalty of $2,000 per ton of SO_2 and required to reduce an additional ton of pollution the next year for each ton of excess pollutant emitted.

Clean Coal Technology

The law authorizes a 4-year extension (to Dec.31, 2003) of the Phase 2 compliance deadline for repowering clean coal technology, and exempts DOE clean coal projects from the NSPS and the major modification provisions of the Clean Air Act.

Studies and Reports

Title IV includes a number of studies and reports focused on acid rain issues, including a national acid lakes registry, reports on monitoring of acid rain in Canada, studies of high altitude lakes in Wyoming and of the use of buffering agents in that State, a study of the feasibility of a acid deposition standard, a report on clean coal technology exports, and a special clean coal demonstration project.

In addition, Title IX, Clean Air Research, provides for a continuation of the National Acid Precipitation Assessment Program and for specific studies of acid rain impacts in the Adirondacks and in Western States.

In February, 1995, the EPA released a draft report on the feasibility of a acid deposition standard. The draft study suggests that the mandated reductions under Title IV will provide clear benefits to sensitive surface water, that nitrogen dioxide, as well as sulfur dioxide, plays an important role in acidification of surface waters, and that the allowance trading program is not projected to harm surface waters.

Chapter 3

ACID RAIN, AIR POLLUTION, AND FOREST DECLINE

Adela Backiel

INTRODUCTION

Certain tree species in several areas of the United States are showing evidence of a decline in productivity, many from unknown causes. Acid rain and other air pollutants are seen as possible causes, or contributing factors, in these declines. The primary pollutants under investigation are sulfur dioxide, nitrogen oxides, reactive hydrocarbons, and ozone. The effect of acid rain and air pollution on trees and forests, both confirmed and disputed, plays a role in the legislative debate over pollutant emission controls.

Few tree declines in the United States can be conclusively attributed to air pollution. The most common hypothesis for declines is one of general stress where natural stresses such as insects or diseases work in concert with one or more anthropogenic factors such as air pollution to cause a decline in productivity. Many other hypotheses exist, including both direct effects of air pollutants on tree foliage and indirect effects from pollution-induced changes in the soil.

Federal research on the effects of acid deposition and air pollution on forests has increased significantly over the past few years. The current forest-related research program is headed by the Forest Service and the Environmental Protection Agency (EPA). The key forestry issues relating to the debate on pollutant emission controls are whether the problem is so exigent as to warrant immediate controls, and if controls are warranted, which pollutants the program should address. Opponents to pollution control bills state that additional regulation should be postponed until these and other questions can be conclusively answered. Advocates of additional pollution controls contend that delay could cause irreversible and irrevocable damage.

In 1980, an independent, 10-year organization was created, the National Acid Precipitation Assessment Program (NAPAP), to coordinate acid rain research activities and periodically report to Congress on all aspects of the acid rain issue. The NAPAP research program reports expenditures of over $80 million annually. The final report of the 10-year National Acid Precipitation Assessment Program (NAPAP) is now in draft form.

In July 1989, the Bush Administration, having pledged to "protect the air we breathe," submitted a proposal to amend the Clean Air Act (H.R. 3030/S. 1490). The congressional debate has shifted from whether to attempt to curb acid rain to what measures are appropriate to do so. Title V of the legislation addresses acid rain and sets goals for SO_2 and NO_x emissions. H.R. 3030 passed the House May 23, 1990. S. 1630, which combines several bills along with elements from the Administration's proposal, was reported from the Committee on Environment and Natural Resources Dec. 20, 1989. A compromise amendment on the reported bill, worked out by Administration officials and ten Members of the Senate, was debated on the Senate floor -- along with other amendments -- and passed Apr. 3, 1990. H.R. 3030 and S. 1630 are in conference.

ISSUE DEFINITION

Trees in several areas of the United States are showing evidence of decline. Although U.S. forests have periodically suffered ill health from known causes, there is little scientific consensus explaining the cause and effect relationship for many of the current declines. Acid rain and air pollution generally are suspected agents in some of the declines, but a wide range of interconnected causes may be involved.

This issue brief discusses the nature and extent of forest decline, some explanations for this decline, and the Federal response to address this situation. The central issue is whether acid rain and/or air pollution contributes to the cause of forest decline, and how this factors into the current congressional debate over pollutant emission controls.

BACKGROUND AND ANALYSIS

Forest decline involves a complex set of biological and nonbiological stress factors, as opposed to a plant disease in which death of one species can be attributed to one cause. Symptoms of decline can be obvious or invisible to the naked eye, rapid or progressive, chronic or acute. Resulting damage can include reduced growth, crown and twig dieback, yellowing of leaves and needles, decreased leaf size, thinning and death of leaves and needles, and deterioration of roots.

Most of the attention and research on the possible air pollution effects on forests have been focused on the eastern and southern parts of the United States. Despite documented ozone damage in southern California for Ponderosa and Jeffrey pines, California black oak, and white fir, and despite specific point source pollution problems (e.g., forested areas surrounding smelters), the West has generally been considered less threatened by regional air pollution.

In scattered areas of the Appalachian and Blue Ridge Mountains, damage to eastern white pine has been confirmed. At high elevations, red spruce has also suffered significantly increased mortality and growth reductions for the past 20 to 25 years, but without conclusive links to air pollutants. Loss of foliage and fine roots are the most common visible symptoms, sometimes leading to more susceptibility to damage from insects and disease. Red spruce at low elevations have also experienced growth reductions but without other obvious symptoms.

Forest surveys for the Piedmont regions of Alabama, Georgia, North Carolina, and South Carolina have indicated unexpected growth reductions of approximately 20% for loblolly and shortleaf pines during the last 10 years. An unexpected decrease in rates of tree growth of longleaf and slash pines has also occurred in the coastal plain regions of other southern states.

In the Ohio River valley, white ash, tulip poplar, sugar maple, red oak, white oak and various bark lichens show various symptoms and degrees of damage from the pollution. The health of lichens is often used as an indicator of pollution in a given area.

Forest decline is best understood within the context of the entire forest ecosystem. Generally, the ecosystem has the capability to adjust to change and maintain an equilibrium; if any one of the ecosystem components is substantially altered, as from a natural or man-made stress factor, the entire ecosystem can be affected.

Most forest stresses are familiar, natural substances or situations, encountered and accommodated on a daily basis, which can be cast into three categories: biological, physical, and chemical factors. Biological factors primarily include insects, fungi, and viruses. Physical factors can involve climatic extremes of heat and cold, moisture supply, and high winds; geographic location; age of the forest; wildfires; and human activities such as prescribed burning, logging, or construction damage such as soil compaction. Chemical factors include the natural and anthropogenic (human-origin) gases such as sulfur dioxide, nitrogen, and ozone; potentially toxic metals such as aluminum and lead; acidic substances such as sulfuric and nitric acids; and growth-altering organic chemicals, such as ethylene.

Because of the resiliency of the forest ecosystem, most trees and forests recover from the effects of stress. However, at a critical time, one factor could -- like the straw that broke the camel's back -- push a tree, or an ecosystem, from a weak but stable situation, into a vulnerable and declining condition. Also, each individual tree species and/or forest can respond to these stresses in a different manner. At the same time, various stresses or declines can produce the same symptoms.

AIR POLLUTANTS AS A SOURCE OF FOREST STRESS

One of the most recent types of stress that trees and forests have had to accommodate is air pollution. Pollutants are generally identified as primary or secondary, depending on how they are formed. Primary pollutants are directly emitted into the atmosphere, predominantly in gaseous forms and include sulfur dioxide (SO_2), nitrogen oxides (NO_x), and reactive hydrocarbons (RHC), which are carbon compounds and mixtures such as ethylene and gasoline. The burning of fossil fuels -- coal, natural gas, and oil -- produces a large portion of these pollutants; industrial, utility, and power plants emit large quantities of the SO_2 and NO_x. Mobile sources, such as transportation vehicles, are responsible for large amounts of NO_x and RHCs. Petroleum refining and storage produces reactive hydrocarbons, although the majority are produced from numerous smaller sources such as the use of paint solvents and dry cleaners. Metal ore smelters can release SO_2, in addition to toxic elements such as lead, cadmium, nickel, and fluoride. Lead can also come from burning leaded gasoline.

Secondary pollutants are formed during chemical reactions in the atmosphere involving the primary pollutants. One of the most commonly known is acidic deposition, which is produced when SO_2 and sulfur oxides (SO_x) or NO_x combines with oxygen in the air to form acidic gases or particulates (dry deposition); in combination with moisture, these processes can cause acid snow, hail, dew, fog, or rain (wet deposition, or "acid rain").

Another type of secondary air pollutant -- photochemical oxidants -- is produced when energy provided by sunlight triggers reactions involving NO_x and atmospheric oxygen. Ozone (O3), the most common oxidant, is a gas that is extremely toxic to both plants and humans. Like primary pollutants, secondary pollutants can travel long distances. Secondary pollutants can best be controlled by limiting their precursors, or by restricting other chemical compounds that may control the precursor rate of transformation. For example, production of ozone is highly dependent on the ratio of hydrocarbons to NO_x in the atmosphere -- thus control of ozone could theoretically be accomplished by limiting either hydrocarbons or NO_x to meet the correct ratio, though NO_x can be controlled more easily than hydrocarbons.

HYPOTHESES TO EXPLAIN THE ROLE OF AIR POLLUTANTS IN FOREST DECLINE

Many studies have shown that local (point source) air pollution can harm vegetation including forest growth. However, the effects of regional air pollution on forest growth are not as well documented. Regional air pollutants may have direct effects on trees through the leaves and stems, and indirect effects, primarily through the soil.

Many forest scientists also believe that air pollution could make a tree or forest more susceptible to natural stresses, such as insects, diseases, drought, or winter frost damage. Consequently, the most common theory advanced to explain the cause of current forest declines, called the general stress hypothesis, is that varying combinations of climate extremes, pests and pathogens, or other natural stresses, in concert with one or more anthropogenic factors, may be responsible. Very few of the American forest declines can be conclusively attributed, in whole or in part, to air pollution.

Many hypotheses exist in addition to this general stress hypothesis. These include both direct effects of air pollutants on tree foliage and indirect effects from pollution-induced changes in the soil. Appendix 1 summarizes the leading hypotheses derived from a survey of scientific literature. Possible U.S. forests of concern are also indicated, although the effects are likely to vary with the tree species, soils, climate, and the variety and concentration of pollutants found at any individual site.

The following general points can be derived from reviewing Appendix 1 regarding the link between air pollutants and forest decline: (1) the wet deposition of acidic solutions of sulfur and nitrogen oxides (commonly referred to as acid rain) is only one of several possible forest damaging effects of pollution; (2) "dry" sulfur dioxide may enter tree leaves and then react with cell moisture to form acids; (3) several hypotheses do not involve any acidic effects from pollution; and (4) the hypotheses are not mutually exclusive.

FEDERAL RESEARCH ACTIVITIES

The USDA Forest Service has been researching certain forest declines and possible effects of air pollution on forests for many years. In 1980, an independent, 10-year organization was created, the National Acid Precipitation Assessment Program (NAPAP), to coordinate acid rain research activities and periodically report to Congress on all aspects of the acid rain issue. The NAPAP research program reports expenditures of over $80 million annually.

The final report of the 10-year National Acid Precipitation Assessment Program (NAPAP) is now in draft form. Report #16, "Changes in Forest Health and Productivity in the USA," reaches five conclusions about forest health and air quality:

"(1) The vast majority of forests in the United States and Canada are not affected by decline (see Section 1.7 and 11.1 for definition of decline). (2) There is experimental evidence that acidic deposition and associated pollutants can alter the resistance of red spruce to winter injury; through this mechanism, acidic deposition may have contributed to dieback and mortality of red spruce at high elevations in the northern Appalachians. (3) Natural stresses are important factors contributing to recent declines of sugar maples. (4) Natural stresses are important factors contributing to growth reductions in natural pine stands in the Southeast. (5) Ozone is an important factor in a decline of pines in southern California and is the pollutant of greatest concern with respect to possible regional scale impacts on North American forests."

The Terrestrial Effects Task Group of NAPAP is a multi agency effort to explore the effects of acid deposition on crops and forests. The majority of work focuses on forest effects. The NAPAP research budget for forestry research has increased significantly over the past few years from about $1 million in FY84 to over $19 million in FY89. Most of the forest related research coordinated through NAPAP is supported by the Forest Service and the Environmental Protection Agency (EPA) in an organization called the Federal Management Group. This research is conducted through four, 5-year Cooperatives focused on specific forest types. The National Council of the Paper Industry for Air and Stream Improvement (NCASI), the research arm of the forest products industry, and scientists from many universities also contribute to the research Cooperatives. In support of the Cooperatives, a Deposition Monitoring Program is managed by EPA as part of the Federal Management Group.

The current Cooperatives address four forest types: the southern pine forests, the spruce-fir forests, western coniferous forests, and the eastern hardwood forests. The Cooperatives focus on determining the effects of acid deposition or air pollution on trees and forests and whether they contribute to declines in forest productivity. These Cooperatives will likely continue to focus Forest Service and EPA research, even with the conclusion of the NAPAP Program.

POLICY AND LEGISLATIVE ISSUES

Throughout the legislative debate on acid rain, issues have focused on what control programs, if any, are appropriate to abate acid rain. A continuing question has been: is the problem so exigent as to warrant immediate controls? Some have argued that delaying action would allow better information on the problem and solutions which might result in a more effective and less costly program. Others contend that delay risks irreversible and unacceptable damage.

An understanding of the effects of acid rain and air pollutants on various resources is central to the debate over additional regulations to control air pollutants. With regard to forestry, there are still many unknowns as to the extent of damage from air pollutants, and the cause of productivity declines and their impacts on the forest products industry.

The Reagan Administration took the position that additional regulation was premature and more research into the effects of air pollution was necessary. A joint statement on transboundary air pollution issues was issued in January 1986 from the Special Envoys appointed by President Reagan and Canadian Prime Minister Mulroney. Their recommendation on forest effects was that "research should be accelerated to investigate the potential link between forest and tree decline and acid rain as a causal or contributing factor." The increased current research program is partially the result of this recommendation. As stated in the Joint Report to the Bilateral Advisory and Consultative Group on the Status of Canadian/U.S. Research in Acidic Deposition in Feb.25, 1987, the current research program expects some answers in the short-term as to the role of acid deposition in observed forest effects. The program is also expected, over the longer term, to establish the effects on forests, if any, of acid rain, air pollution, and natural stresses, alone or in various combinations.

In July 1989, the Bush Administration, having pledged to "protect the air we breathe," submitted a proposal to amend the Clean Air Act (H.R. 3030/S. 1490). The congressional debate has shifted from whether to attempt to curb acid rain to what measures are appropriate to do so. Title V of the legislation addresses acid rain and sets goals for SO_2 and NO_x emissions. H.R. 3030 passed the House May 23, 1990. S. 1630, which combines several bills along with elements from the Administration's proposal, was reported from the Committee on Environment and Natural Resources Dec. 20, 1989. A compromise amendment on the reported bill, worked out by Administration officials and ten Members of the Senate, was debated on the Senate floor -- along with other amendments -- and passed Apr. 3, 1990. H.R. 3030 and S. 1630 are in conference.

Although the lack of undisputed evidence does not enable the scientific community to speak with one voice, empirical research, circumstantial evidence, and some of the best available scientific information suggest that air pollutants could be working alone or in conjunction with natural stresses to damage forest health, vigor and productivity. While the potential benefits from SO_2 and NO_x reductions could be substantial, such reductions alone may not adequately protect ecosystems, such as forests.

CONGRESSIONAL HEARINGS, REPORTS AND DOCUMENTS

U.S. Congress. House. Committee on Agriculture. Forest Ecosystems and Atmospheric Pollution Research Act of 1986; report to accompany H.R. 2631.Washington, U.S. Govt. Print. Off., 1986. (99th Congress, 1st session. House Report no.99-776)

U.S. Congress. House. Committee on Agriculture. Subcommittee on Forests, Family Farms, and Energy. Hearing, 99th Congress, 2d session, on H.R. 2631 and H.R. 2963. May 13, 1986. Washington, U.S. Govt. Print. Off., 1986."Serial no.99-27"

U.S. Congress. House. Committee on Interior and Insular Affairs. Authorizing and directing the Secretary of Agriculture to engage in a ten-year research program; report to accompany H.R. 2963. Washington, U.S. Govt. Print. Off., 1985. (99th Congress, 1st session. House. Report no.99-344, Part 1)

POLLUTANT	HYPOTHESIS	EXPECTED SYMPTOMS	POSS. FORESTS AFFECTED
Deposition of wet and dry acidic sulfur and nitrogen oxides (Sox and Nox)	Acid deposition may reduce normal nutrient level available to trees. 1) It can speed the normal leaching of nutrients directly from leaves or needles. If affected trees cannot sufficiently replace these nutrients, deficiencies can alter normal life and growth processes.	Yellowing and browning of foliage	Forests that are frequently cloud or mist-covered and exposed to atmospheric pollution
	2) It can acidify forest soils. As soils become more acidic, nutrients (such as calcium and magnesium) can be removed.	Soil acidification Yellowing and browning of foliage	Forests on nutrient-poor soils
Fertilization by Nitrogen Oxides (NO_x)	1) Too much nitrogen "fools" trees into continuing to grow late in the fall. Winter cold can damage and kill foliage that is not prepared.	Red-brown discoloration of preceding year's foliage	High-elevation forests; e.g., red spruce in northeastern Appalachia
	2) Nitrogen-fertilized plants respond by increasing growth above ground, but this is not offset by increased root growth. Total root system may actually be decreased, making tree more susceptible to drought, other stresses.	Decreased root mass and mycorrhiza	
Gaseous Pollutants Ozone and other Photochemical Oxidants	A sudden onset or sharp rise in ozone can damage or kill leaf cells, leading to decreased photosynthesis, loss of water, and increased leaching or removal of nutrients from foliage.	Death of cells in leaves and needles Purple/black on broad leaves Premature aging of foliage Reduced growth	+White fir, California black oak, Ponderosa & Jeffrey Pines in southern California +White pine, eastern U.S. Loblolly and shortleaf pine in Piedmont forests of AL, GA, NC, SC
Sulfur Dioxide (SO_2)	Chronic Ozone exposure in combination with acidic rain or fog increased nutrient loss from foliage, which leads to reduced photosynthesis and root growth SO_2 can enter plant through stomata and react with water to form acids. Acute or chronic exposure can damage or kill cells in leaves, reducing photosynthesis.	Yellowing or browning of foliage Yellowing or browning of foliage Often no visible injury Premature needle loss	High-elevation forest in Northeast Point-source problems (e.g., smelters in Sudbury, Ontario)
Trace Metals Aluminum Mobility	As soil acidity rises, aluminum increases to toxic levels and can displace beneficial nutrients, inhibit their uptake, and damage fine root hairs, leading to nutrient and moisture stress.	Loss of leaves or needles Soil acidification Death of fine roots	No evidence in U.S. (highly suspected in W. German forests)
Heavy Metal Deposition	Metals exist in soils in trace amounts. Excess amounts could interfere with photosynthesis, root growth, and reproduction.	Slow rate of decomposition	High-elevation spruce-fir ecosystems in southeastern U.S. Localized pollution problem areas particularly close to smelters
Growth-Altering Substances	Synthetic organic substances may stimulate abnormal growth and development, particularly in foliage.	Browning and death of leaves and needles Loss of green foliage	+Loblolly pine in North Carolina

+ Denotes general scientific acceptance of linkage to noted causal agents.

BIBLIOGRAPHY WITH ABSTRACTS

Abate, T.
Swedish scientists take acid-rain research to developing nations.
Bioscience; Dec 31 1995. Description: In the realm of acid-rain research, Sweden looms large on the world stage. It is the country where scientists first proved more than 30 years ago that airborne chemicals could and did cross international boundaries to acidify lakes and forests far from where the pollution was generated. Now, Swedish scientists are leading an international effort to map acid-rain patterns in the developing countries of Asia, where new industrial activity seems to be recreating problems that European and North American policy makers have already taken steps to solve. Topics covered in this article include acid rain on the rise in Asia; visualizing and validating the data; funding as the key to steady research. Vol. 45; no. 11; pp. 738-740

Abelson, P.H. Clean coal technology. Science (Washington, D.C.); Dec 07 1990. Description: One of the major technology challenges in the next decade will be to develop means of using coal imaginatively as a source of chemicals and in a more energy-efficient manner. The Clean Air Act will help to diminish the acid rain but will not reduce CO_2 emissions. The Department of Energy (DOE) is fostering many innovations that are likely to have a positive effect on coal usage.
Vol. 2504986; pp. 1317

Acid rain and the NAPAP study
Public Utilities Fortnightly; Mar 01 1990
Description: This article reports on preliminary state of science and technology reports on acidic deposition released by the National Acid Precipitation Assessment Program. These are part of the integrated assessment of acidic deposition and related air pollutants to be released at a later date. SO_2 and NO_x emissions and effects are discussed. Vol. 1255; pp. 51-52

Acid rain revisited. Journal of the Air and Waste Management Association; Jun 30 1990. Description: This article reviews calculations of the estimates reported in this Newletter in 1983 on the contributions of nitric and sulfuric acids to deposition that may originate with motor vehicles. Previous estimates can now be updated, based on the recently released 1985 emissions inventory that was compiled by the National Acid Precipitation Assessment Program (NAPAP). This inventory relied in turn on the US Environmental Protection Agency's National Emissions Data System (NEDS), the 1985 version of which is the latest containing both point source and area source information in sufficient detail to be useful for this purpose.
Vol. 406; pp. 942

Adams, M.B.; O'Neal, E.G.
Effects of ozone and acidic deposition on carbon allocation and mycorrhizal colonization of Pinus taeda L. seedlings
Forest Science; Mar 31 1991
Description: Patterns of carbon allocation

and mycorrhizal colonization were examined in loblolly pine seedlings from two half-sib families exposed to three ozone treatments (charcoal-filtered air, ambient air + 80 ppb O_3, and ambient air + 160 ppb O_3) and three rain pH levels (5.2, 4.5, and 3.3) for 12 weeks in open-topped chambers in a field setting. No statistically significant effects of ozone or rain pH were detected on biomass, root:shoot ratios, or carbon allocation; some consistent patterns were observed, however.
Vol. 371; pp. 5-16

Albers, P.H.; Camardese, M.B.
Effects of acidification of metal accumulation by aquatic plants and invertebrates. 2. Wetlands, ponds and small lakes. Environmental Toxicology and Chemistry; Jun 30 1993 Description: Compared were concentrations of Al, Cd, Ca, Cu, Fe, Pb, Mg, Mn, Hg, Ni, P, and Zn in water, plants, and aquatic invertebrates of wetlands, ponds, and small lakes in Maryland and Maine. The accumulation of metals by aquatic plants and insects and the concentration of metals in water were not greatly affected by pH. None of the metal concentrations in water significantly correlated with metals in insects.
Vol. 126; pp. 969-976

Albers, P.H.; Camardese, M.B.
Effects of acidification on metal accumulation by aquatic plants and invertebrates. 1. Constructed wetlands. Environmental Toxicology and Chemistry; Jun 30 1993 Description: Compared were concentrations of Al, Cd, Ca, Cu, Fe, Hg, Pb, Mg, Mn, Ni, P, and Zn in water, plants and aquatic insects of three acidified (pH [approximately] 5.0) and three nonacidified (pH [approximately] 6.5) constructed wetlands. Concentrations of Zn in water and bur-reed (Sparganium americanum) were higher in acidified wetlands than in nonacidified wetlands. Floating nonrooted plants contained mean concentrations of Fe, Mg, and Mn that were higher than recommended maximum levels for poultry feed.
Vol. 126; pp. 959-967

Albritton, D.L.; Fehsenfeld, F.C.; Tuck, A.F.
Instrumental requirements for global atmospheric chemistry
Science (Washington, D.C.); Oct 05 1990
Description: The field of atmospheric chemistry is data-limited, primarily because of the challenge of measuring the key chemical constituents in the global environment. Several recent advances, however, in rugged, portable, remote-sensing, ground-based instrumentation and accurate, fast-response airborne instrumentation have provided powerful tools for the understanding of stratospheric ozone, particularly in polar regions. Current discoveries of the role of heterogeneous chemical processes point to the need for better techniques for characterization of stratospheric aerosols
Vol. 2504977; pp. 75-81

Alewell, C; Matzner, E.; Bredemeier, M; Blanch, K.
Soil solution response to experimentally reduced acid deposition in a forest ecosystem. Journal of Environmental Quality; May 31 1997
Description: In order to measure and predict reversibility of soil solution acidification under experimentally reduced acid input, a manipulation study with artificial 'preindustrial' throughfall was established. A roof was installed underneath the canopy in a Norway Spruce stand of the German Soiling area. Water failing onto the roof was adjusted to clean rain concentrations before redistribution. Soil solutions were collected with suction cup lysimeters at various depths and were analyzed for major ions. The response of soil solution chemistry in the upper soil (10 cm depth) to a reduction of N, SO_4, and H input was rapid. While NO_3 concentration in deeper soil layers reached input levels after 2 yr of treatment, SO_4 concentration in the seepage water at 1 m depth remained high relative to the reduced input due to a release of formerly stored S from the soil.
Vol. 26; no. 3; pp. 658-665

Alkezweeny, A.J.
Field observations of in-cloud nucleation and the modification of atmospheric

aerosol size distributions after cloud evaporation Journal of Applied Meteorology; Dec 31 1995 Description: The authors measured aerosol and droplet size distributions in the range from 0.1 to 50 {mu}m, concentrations and sizes of precipitation particles, concentrations of condensation nuclei, and state parameters in and in the vicinity of a towering summer cumulus cloud. The measurements show that the liquid water content, droplet concentration, and vertical velocity all peaked in the upper half of the cloud, but the droplet mean diameter increased with altitude. Vol. 34; no. 12; pp. 2649-2654

Allen, Robert.B.; Clinton, Peter.W.; Davis, Murray.R.
Cation storage and availability along a Nothofagus forest development sequence in New Zealand
Canadian Journal of Forest Research; Mar 01 1997 Description: Soil cations and pH were determined in relation to the development of even-aged (10, 25, 120, and > >150-year-old stands) mountain beech (Nothofagus solandri var. cliffortioides (Hook.f.) Poole) forest after catastrophic canopy disturbance. Live stem biomass varied from 1 to 273 Mg/ha between the seedling (10 years) and pole (120 years) stages, respectively, but was less in the mature stage (> >150 years; 245 Mg/ha). Vol. 27; no. 3; pp. 323-330

Allison, J.E.
Fortuitous consequence: The domestic politics of the 1991 Canada-United States agreement on air quality
Policy Studies Journal; Jul 01 1999 Description: Following more than a decade of negotiations, the Canada-United States Agreement on Air Quality entered into force on March 13, 1991, with the signatures of then-Canadian Prime Minister Brian Mulroney and US President George Bush. Why was it so difficult for Canadian and US negotiators to reach agreement? The author argues that Canadian and US domestic politics were the primary impediments to resolving the US-Canada acid rain dispute.
Vol. 27; no. 2; pp. 347-359

Alm, L.R.
Regional influences and environmental policymaking: A study of acid rain
Policy Studies Review; Jan 31 1993 Description: This study uses a four-region classification of states to investigate the impact of regionalism on the acid rain policy debate. Using multiple regression analyses and controlling for such factors as total pollution emissions, coal production, and precipitation pH, it was found that a state's activity level with respect to involvement in the acid rain policy debate was significantly influenced by both its commitment to environmental protection and its regional differentiation.
Vol. 214; pp. 638-650

Alm, L.R.; Davis, C.
Agenda setting and acid precipitation in the United States. Environmental Management; Jan 31 1993. Description: The objective of this research was to analyze the retention of acid precipitation as a viable policy issue on the Congressional agenda during the 1980s. Issue maintenance (a term borrowed from Barbara Nelson's discussion of the four stages associated with agenda decision making) was examined in relation to a set of issue characteristics originally developed by R. Cobb and C. Elder, i.e., concreteness, social significance, temporal relevance, complexity, and categorical precedence. Vol. 176; pp. 807-816

Amokrane, H.; Saboni, A.; Caussade, B.
Experimental study and parameterization of gas absorption by water drops
AIChE Journal (American Institute of Chemical Engineers); Dec 31 1994 Description: Mass transfer between liquid drops and a continuous gas phase occurs in a large number of industrial processes and many engineering disciplines such as chemical and nuclear engineering, atmospheric sciences, environmental engineering, and so on. Liquid-phase mass-transfer coefficients are determined for the absorption of sulfur dioxide by water drops larger than 1.1 mm in dia.
Vol. 4012; pp. 1950-1960

Anderson, P.D.; Houpis, J.L.J.
  Foliar nutrient status of Pinus ponderosa exposed to ozone and acid rain
  Plant Physiology, Supplement; May 31 1991 Description: A direct effect of foliar exposure to acid rain may be increased leaching of nutrient elements. Ozone exposure, through degradation of the cuticle and cellular membranes, may also result in increased nutrient leaching. To test these hypotheses, the foliar concentrations of 13 nutrient elements were monitored for mature branches of three clones of Pinus ponderosa exposed to ozone and/or acid rain. The three clones represented three distinct levels of phenotypic vigor. Vol. 961; pp. 173

Aoki, Ichiz; Kikkawa, Yoshitsugi
  Technical efforts focus on cutting LNG plant costs. Oil and Gas Journal; Jul 03 1995. Description: LNG demand is growing due to the nuclear setback and environmental issues spurred by concern about the greenhouse effect and acid rain, especially in the Far East. However, LNG is expensive compared with other energy sources. Efforts continue to minimize capital and operating costs and to increase LNG plant availability and safety. Technical trends in the LNG industry aim at reducing plant costs in pursuit of a competitive LNG price on an energy value basis against the oil price.
  Vol. 93; no. 27; pp. 43-47

Aono, Hiromichi; Sugimoto, Eisuke; Mori, Yoshiaki; Okajima, Yasuhiro
  Electromotive force responses of Cl_2 gas sensor using $BaCl_2$-KCl solid electrolyte Journal of the Electrochemical Society; Nov 30 1993. Description: Chlorine is the most important halogen in industrial production. Chlorine exhaust gas has become a serious problem with regard to air pollution and acid rain in recent years. Solid electrolyte-type gas sensors are superior for SO_x or CO_2 detection because of their rapid response.
  Vol. 14011; pp. 3199-3203

Appanna, V.D; Huang, J; St. Pierre, M.
  Cesium stress and adaptation in pseudomonas fluorescens
  Bulletin of Environmental Contamination and Toxicology; May 31 1996.
  Description: Industrialization and acid rain have led to a marked increment on the bioavailability of numerous metals. These metallic pollutants pose a serious threat to the ecosystem due to their ability to interact negatively with living organisms. Thus, considerable effort has been directed towards the development of environmentally-friendly technologies tailored to the management of metal wastes. As microbes are known to adapt to most environmental stresses, they constitute organisms of choice in the study of molecular adaptation processes. The adaptive features may be subsequently engineered for biotechnological applications. Vol. 56; no. 5; pp. 833-838

Appleton, E.L
  Cross-media approach to saving the Chesapeake Bay. Environmental Science and Technology; Dec 31 1995.
  Description: A project EPA began in August will investigate the possibility of cross-media emissions trading as a new approach to reducing nitrogen loadings to the Chesapeake Bay. Working with the Environmental Defense Fund (EDF), the Agency hopes to device a NO_x trading framework along the lines of existing sulfur dioxide trading plans to control acid rain. The Chesapeake Air Project will examine the feasibility of using emissions trading between and water sources, including trading credits between power plants and mobile sources, to reduce the atmospheric deposition of nitrogen to the bay. Vol. 29; no. 12; pp. 550A-555A

Aris, R.; Christian, D.; Sheppard, D.; Balmes, J.R.
  Acid fog-induced bronchoconstriction. The role of hydroxymethanesulfonic acid American Review of Respiratory Disease (New York); Mar 31 1990
  Description: Hydroxymethanesulfonate (HMSA), the bisulfite (HSO3-) adduct of formaldehyde (CH2O), is a common constituent of California acid fogs. HMSA, most stable in a fog pH range of 3 to 5,

dissociates at 6.6, the pH of the fluid lining human airways. The dissociation of inhaled HMSA should theoretically generate sulfur dioxide and CH2O, both of which have bronchoconstrictor potential. Thus, we hypothesized that HMSA may have a specific bronchoconstrictor effect independent of its strength as an acid.
Vol. 1413; pp. 546-551

Asai, Ei-Ichiro; Hata, Kunihiko; Futai, Kazuyoshi
Effect of simulated acid rain on the occurrence of Lophodermium on Japanese black pine needles
Mycological Research; Nov 01 1998
Vol. 102; no. 11; pp. 1316-1318

Asolekar, S.R.; Valentine, R.L.; Schnoor, J.L.
Kinetics of chemical weathering in B horizon spodosol fraction
Water Resources Research; Apr 30 1991
Description: In response to acidic deposition a variety of physico-chemical processes occur. Studies on a B horizon soil from Maine have been conducted to determine the weathering rate dependence on hydrogen ion concentration in soil solution. Effects of soil concentration and solution chemistry on chemical weathering rate were also investigated.
Vol. 274; pp. 527-532

Ayers, G.P.; Gillett, R.W.
Tropospheric chemical composition - Overview of experimental methods in measurement. Reviews of Geophysics and Space Physics; Aug 31 1990. Description: Methods used to measure various aspects of tropospheric chemical composition are discussed. Particular attention is given to the determination of trace species at levels which can be below 1 part in 10 to the 12th in the gas, aqueous, and aerosol phases. It is shown that the precision of a measurement tends to be an inverse function of atmospheric mixing ratio and species reactivity. Vol. 28; pp. 297-314

Baedecker, P.A.; Reddy, M.M.
The erosion of carbonate stone by acid rain: Laboratory and field investigations
Journal of Chemical Education; Feb 28 1993. Description: This paper describes a laboratory experiment on the effects of acidic deposition on carbonate stone erosion. It can serve as the basis for an undergraduate (or pre college) experiment in environmental chemistry. Recent field investigations are described that provide measurements of carbonate stone dissolution and mechanical erosion under weathering conditions that are prevalent in the eastern US. Vol. 702; pp. 104-108

Bandyopadhyay, J.K; Annamalai, S; Gauri, K.L.
Application of artificial neural networks in modeling limestone-SO_2 reaction
AIChE Journal; Aug 31 1996. Description: Four varieties of limestone distinguished on the basis of pore-size distributions were exposed in dynamic 10-, 20- and 50-ppm SO_2 atmospheres for up to 500 h at 25 C and 100% rh. The resulting conversion of the limestone was measured as a function of the reaction product formed and the change in porosity. These conversions could be predicted correctly using either the shrinking unreacted core model or the distributed pore model. An artificial neural network (ANN) was also trained for the purpose. All three approaches predicted conversions that fitted well with the observed data; however, those predicted by ANN were the most accurate.
Vol. 42; no. 8; pp. 2295-2302

Banta, H.M.; Bierman, S.L.
Acid rain legislation's complex problem - Fair and efficient emissions limitation
Public Utilities Fortnightly; Mar 01 1990
Description: The writers of this article take a gloomier view of the potential of an emissions reduction compliance market. It cannot be relied upon to even out disparities in the costs which different utility systems will have to incur, they say - i.e., promote fairness - and it will have the added disadvantage that it will work against the entry of nonutility electricity generators into this field. The monopolistic and noncompetitive nature of the electric utility industry is seen as a major problem - one which will preclude operation of an emissions market. Vol. 1255; pp. 19-21

Banwart, W.L.; Finke, R.L.; Porter, P.M.; Hassett, J.J.
Sensitivity of twenty soybean cultivars to simulated acid rain.
Journal of Environmental Quality; Jan 31 1990. Description: Efforts to assess the effect of acid rain on soybean (Glycine max (L.) Merr.) have shown variable results. Simulated acid rain has been reported to cause increases, decreases, and no significant effect on yield. Although few parameters were identical among the diverse studies reported in the literature, one common difference was the choice of cultivar. In this study, 20 soybean cultivars were screened to determine their relative sensitivity to simulated acid rain. Soybean was grown in 1984, 1985, and 1986 in field plots in east central Illinois. Plots were protected from ambient rain and treated twice weekly with simulated rain of pH 5.6 (control) or pH 3.0. Early in each growing season visible leaf injury was noted for all 20 cultivars, and level of injury was significantly higher for plants receiving the more acidic treatment (pH 3.0). Average yield for the 20 cultivars was approximately the same when plants were treated with simulated rain of pH 3.0 as when plants were treated with stimulated rain of pH 5.6. Vol. 192; pp. 339-346

Banwart, W.L.; Ziegler, E.L.; Porter, P.M.
Field corn response to acid rain-drought stress interaction.
Journal of Environmental Quality; Jan 31 1990. Description: Two studies were conducted in 1988 to examine the effects of simulated acid rain in combination with various levels of drought stress on the grain yield of field grown corn (Zea mays L., B73 x Mol17 and FS854). In both studies corn was treated with twice weekly applications of simulated rainfall of pH 5.6 or 3.0 at amounts that totaled 100% (30 cm), 50% (15 cm), and 25% (7.5 cm) of the seasonal average for Champaign-Urbana, Il. In addition to those treatments, in one of the studies the plants were subjected to daily wetting with the appropriate simulated rain from tassel emergence through pollination and fertilization. In both studies, reduced moisture levels resulted in significant reduction in grain yield but simulated rain of pH 3.0 had no effect on yield at any of the moisture levels studied.
Vol. 192; pp. 321-324

Bartels, C.W.
Allowance trading made easy: The cash-forward settlement. Electricity Journal; Nov 30 1993. Description: Centralized trading for cash-forward settlement adopts the most useful aspects of traditional commodity trading of spot cash and futures contracts, but eliminates those aspect which are most problematic for the regulated utility industry. Title IV of the Clean Air Act Amendments of 1990 created a system of tradeable pollution rights which promises to achieve U.S. acid rain reduction goals at a savings of billions of dollars compared to traditional means of environmental regulation. The ultimate success of this allowance trading program depends in large part on the utility industry's capacity to integrate a new type of transaction into planning, decision-making, and cost recovery mechanism.
Vol. 69; pp. 71-75

Barth, M.C.
Numerical modeling of sulfur and nitrogen chemistry in a narrow cold-frontal rainband: The impact of meterological and chemical parameters
Journal of Applied Meteorology; Jul 31 1994. Description: To better understand the impact of various meteorological and chemical parameters on chemical deposition from winter storms, the chemistry and microphysics of a narrow cold-frontal rainband and its associated stratiform region were examined with a two-dimensional numerical cloud model. The peak precipitation was associated with the lifting at the leading edge of the cold front. Vol. 33; no. 7; pp. 855-868

Basman, A.R.; Chorley, G.B.; Adeloju, S.B.
Estimating the corrosion rate of mild steel in sulfuric acid by a hydrogen evolution method. Journal of Chemical Education; Mar 31 1993. Description: An experiment was designed to demonstrate the

connection between the anodic reaction (oxidation of a metal) and the cathodic reaction (reduction of hydrogen ions), thus enhancing the students' understanding of corrosion as an electrochemical process. Also, treatment of the test results involves using gas laws, which students often find difficult to apply. Unlike the apparatus used in most studies of electrochemical corrosion, the apparatus is very simple to use and does not include complicated electronic equipment.
Vol. 703; pp. 258-259

Bassett, S.M
EPA struggles as 1990 Amendments pass five-year mark. Pollution Engineering; Jan 31 1996. Description: Five years after enactment of the Clean Air Act Amendments of 1990, EPA is taking heat from Congress, industry, environmental groups and some states over its implementation of the statute. While the agency has suffered several major policy setbacks and faces inevitable budgetary cuts, it continues to grapple with a large number of regulations still to be promulgated. Have EPA's actions over the past five years accomplished the goals set forth in the 1990 Amendments? How will the Agency meet future challenges? EPA's performance varies widely depending on the Clean Air Act (CAA) program being considered. Vol. 28; no. 1; pp. 27-28

Bates, T.S.; Lamb, B.K.
Natural sulfur emissions to the atmosphere of the continental United States
Global Biogeochemical Cycles; Dec 31 1992. Description: Seasonal emissions of biogenic sulfur gases from marine and terrestrial sources were assessed for 13 regions of the continental United States. The resulting inventory was compared to volcanic and anthropogenic sulfur emissions from the same regions. Terrestrial biogenic sulfur emissions were less than 1% of the total sulfur emissions in each region. However, during the summer, marine biogenic sulfur emissions may be contributing 20-40% of the total sulfur emissions in western coastal regions.
Vol. 64; pp. 431-435

Batterman, S.
Optimized acid rain abatement strategies using ecological goals
Environmental Management; Jan 31 1992
Description: This article addresses the use of critical loads in optimized emission abatement strategies. Critical loads represent the maximum tolerable deposition possible without adverse impacts, a limit that is highly spatially variable. As deposition targets, critical loads cannot be satisfied at all receptors in Europe. Consequently, there is a need for alternative criteria that still relate to ecological indicators, yet that are feasible, consistent, and equitable.
Vol. 161; pp. 133-141

Baublis, D.C.; Dube, R.J.
Coming out ahead with the Clean Air Act
Proceedings of the American Power Conference; Jan 31 1992
Description: This paper reports that the Clean Air Act (CAA) Amendments of 1990 require utilities to reduce the constituents of acid rain, sulfur dioxide (SO_2), and nitrous oxides (NO_x). Reduction strategies involve modifications to all aspects of the operating system, from the boiler proper and associated equipment, fuel systems, existing pollution control equipment and controls. Millions of dollars will be spent on these modifications. To recoup some of these costs, it is economically beneficial to make performance upgrades at the same outage.
Vol. 542; pp. 1457-1461

Bauer, F.W.
Fully engineered approach to acid rain control
Energy Engineering; Jan 31 1994
Description: Regulations and methods for control of sulfur dioxide and particulate emissions have been developed and are fairly well understood. However, nitrous oxide (NO_x) reduction requirements are only now being mandated. Owners of major emissions sources will be unable to continue to defer compliance with evolving legislation. Vol. 914; pp. 17-32

Baxter, James.W.; Pickett, Steward.T.; Carreiro, Margaret.M.; Dighton, John. Ectomycorrhizal diversity and community structure in oak forest stands exposed to contrasting anthropogenic impacts Canadian Journal of Botany; Jun 01 1999 Description: We compared the ectomycorrhizal community structure of oak forest stands located in either an urban or a rural area. Urban stands had higher N deposition rates, soil heavy metal levels, and earthworm counts than rural stands. Ectomycorrhizal types were quantified on roots of mature oak (*Quercus*) in soil cores and on *Quercus rubra* L. seedlings grown in soil cores in the glasshouse. Vol. 77; no. 6; pp. 771-782

Baylor, J.S. Acid rain impacts on utility plans for plant life extension. Public Utilities Fortnightly; Mar 01 1990. Description: Nearly one-quarter of US generating capacity will be reaching the end of its nominal design life in the 1990s. Electric utilities are constrained by costs and uncertainties associated with pending air cleanup legislation to undertake restoration or replacement of many units' components in order to extend their useful lives significantly. However, as this article points out, a recent interpretation by the Environmental Protection Agency of existing law, making plants whose lives are so extended subject to more stringent emission limitations, is likely to impact life extensions with substantially higher costs. Vol. 1255; pp. 22-28

Beamon, J.A.; Linders, M.J. EIA's role in the analysis of the Clean Air Act Amendments of 1990 and the development of the National Allowance Database. Government Information Quarterly; Jan 31 1993. Description: Throughout 1990 the Energy Information Administration (EIA) provided continuous data and analytic support to Congress during its deliberations on Title IV of the Clean Air Act Amendments of 1990 (CAA). Congress requested the Energy Information Administration (EIA) to review and analyze the sections that would affect electric utilities, specifically those relating to acid deposition (Title IV). By providing knowledgeable and impartial analysis, EIA clarified the likely effects of the various legislative proposals and helped Congress finalize the amendments. Even though the CAA is now law, EIA's efforts have not ended. Vol. 101; pp. 25-40

Begley, R. Emissions trading is seen as a new beginning. Chemical Week; Apr 07 1993 Description: Last week's experiment in publicly trading the right to pollute as a commodity on the Chicago Board of Trade may serve as a model for environmental policy. The Environmental Protection Agency required sulfur dioxide (SO_2) emission allowance trading under the acid rain provisions of the 1990 Clean Air Act. The allowance is a new, government-created asset, and people are finding extremely imaginative ways to use them to earn profit, better position their products and their organizations, and clean up the environment, says Rachel Hopp, of counsel with Washington law firm Weinberg, Bergeson, and Neuman and former chief of EPA's acid rain permits and technologies section. Vol. 15213; pp. 8

Bélanger, Luc.; Reed, Austin.; DesGranges, Jean-Luc. Reproductive variables of American black1963-1991. Canadian Journal of Zoology; Jun 01 1998 Description: We examine data from different surveys conducted from 1963 to 1991 in the Baie de l'Isle Verte National Wildlife Area and the surrounding offshore islands, an approximately 20-km2 coastal segment of the St.Lawrence estuary in Quebec. We summarize data regarding various aspects of the nesting ecology of the American black duck (Anas rubripes) (n = 812 nests). Mean laying date, average clutch size, and apparent nesting success did not differ among years. Vol. 76; no. 6; pp. 1165-1173

Bellehumeur, C.; Marcotte, D.; Legendre, P. Estimation of regionalized phenomena by geostatistical methods: lake acidity on the Canadian Shield

Environmental Geology; Jan 18 2000
Description: Abstract This paper describes a geostatistical technique based on conditional simulations to assess confidence intervals of local estimates of lake pH values on the Canadian Shield. This geostatistical approach has been developed to deal with the estimation of phenomena with a spatial autocorrelation structure among observations. It uses the autocorrelation structure to derive minimum-variance unbiased estimates for points that have not been measured, or to estimate average values for new surfaces.

Bencala, K.E.; Zellweger, G.W.; McKnight, D.M. Characterization of transport in an acidic and metal-rich mountain stream based on a lithium tracer injection and simulations of transient storage
Water Resources Research; May 31 1990
Description: Physical parameters characterizing solute transport in the Snake River (an acidic and metal-rich mountain stream near Montezuma, Colorado) were variable along a 5.2-km study reach. Stream cross-sectional area and volumetric inflow each varied by a factor of 3. Because of transient storage, the residence time of injected tracers in the Snake River was longer than would be calculated by consideration of convective travel time alone. Vol. 265; pp. 989-1000

Bennett, R.R.; Barnes, M.W.; Mcdonald, A.J.; Hinshaw, J.C.
Chemical rockets and the environment
Aerospace America; May 31 1991
Description: Recent improvements in the understanding of the earth's atmosphere have prompted an update of past work on the potential environmental effects of chemical propulsion exhaust species. The four general areas of concern are the effects of rocket launches on global warming, acid rain, toxicity, and stratospheric ozone. Based on the current state of knowledge of atmospheric science, it is again concluded that chemical rockets have no significant global effects on the environment. Shuttle liquid and solid rocket motors inject more exhaust products into the atmosphere than any other launcher; however, analyses indicate no significant impact on global environment. Vol. 29; pp. 33-36

Berger, Torsten.W.; Glatzel, Gerhard.
Canopy leaching, dry deposition, and cycling of calcium in Austrian oak stands as a function of calcium availability and distance from a lime quarry
Canadian Journal of Forest Research; Sep 01 1998. Description: Three Austrian oak stands were chosen along a 4-km distance gradient from a lime quarry to study effects of Ca availability both on dry deposition rates and on Ca cycling in these ecosystems. A fourth stand was used as a more regional reference site, some 30 km west of the lime quarry. Calcium bulk precipitation fluxes decreased with increasing distance from the lime quarry, contributing to major differences in available Ca along the transect over the last decades. Vol. 28; no. 9; pp. 1388-1397

Bernow, S; Dougherty, W; Duckworth, M
Quantifying the impacts of a national, tradable renewables portfolio standard
Electricity Journal; May 31 1997
Description: A national renewables portfolio standard with tradable credits, like the one that was proposed in legislation in the 104th Congress, could cost residential consumers just pennies a week and guarantee markets for near-commercial renewable technologies, leading to the development of the renewables industry and multiple environmental benefits in restructured electricity markets. With its inclusion in Rep. Schaefer's Electricity Consumer's Power to Choose Act of 1996, the renewables portfolio standard (RPS) became firmly established as a leading federal mechanism for preserving and promoting the benefits of renewable energy in restructured electricity markets. The RPS policy approach has since consolidated this position through its inclusion in Senator Bumpers's proposed restructuring legislation, thereby acquiring both bi-partisan and bicameral support. Vol. 10; no. 4; pp. 42-52

Bhatti, N.; Streets, D.G.; Foell, W.K.
  Acid rain in Asia
  Environmental Management; Jan 31 1992
  Description: Acid rain has been an issue of great concern in North America and Europe during the past several decades. However, due to the passage of a number of recent regulations, most notably the Clean Air Act in the United States in 1990, there is an emerging perception that the problem in these Western nations is nearing solution. The situation in the developing world, particularly in Asia, is much bleaker. Vol. 164; pp. 541-552

Bi, Shuping
  Acid rain, promoting the accumulation of CO_2 in surface waters. Preprints of Papers, American Chemical Society, Division of Fuel Chemistry; Jan 31 1996
  Description: Acid rain and increase of CO_2 concentration in atmosphere are the important problems to the global changes. In the past, people studied them separately. Someone found that lakes are sources of atmospheric CO_2, but the mechanism is not clear. In this paper, we present our investigation on this question by using chemical equilibrium model. We come to the conclusion that acidified surface waters way hold much more CO_2 than neutral waters. Acid rain, also promotes the accumulation of CO_2 in surface waters. Vol. 41; no. 4; pp. 1355-1359

Bissell, K.; Blake, M.J.; Boucher, P.G.; Gordon, K.; Miller, D.J.; Monk, J.R.; Nelson, S.L.; Norling, N.M.; Sandstrom, D.V.; Urwiller, D.G.; Wilk, G.M.; Worthy, P.M. Question 4: Editorial
  Public Utilities Fortnightly; Nov 08 1990
  Description: This article is a collection of editorials of public service commissioners addressing various issues of concern of the individual chairmen. The topics addressed include a master plan for fiber optic communications for Tennessee, the economic affects of reduced generation assets in a utilities' rate base and alternatives to traditional utility assets, competition between utilities and nonutility generators of electricity, traditional regulation as it applies to the evolving telecommunications and electrical generation industries, the managing of natural resources and energy production to meet human needs and insure our collective prosperity, reorganization and restructuring of the Indiana Utility Regulatory Commission, regulation as a whipping boy being blamed for economic and environmental ills, concerns about affiliate relationships, the effects of acid rain legislation on power generation in North Dakota, accountability to the public, innovative regulation of California's telecommunications utilities as an example for application to other industries, and a call for a national Telecommunications Policy. Vol. 12610; pp. 44-50

Blew, R.D; Edmonds, R.L.
  Precipitation chemistry along an inland transect on the Olympic Peninsula, Washington. Journal of Environmental Quality; Mar 31 1995. Description: The objective of this study was to examine oceanic influences, seasonal variation, and effect of distance from the ocean on the chemistry of bulk precipitation falling on the Pacific coast of Washington State. Bulk precipitation was collected at Sites 4, 13, 24, and 31 km inland from the Pacific Ocean. Mean electrical conductivity of precipitation ranged from 0.47 to 1.02 mS m^{-1} and mean pH ranged from 5.3 to 5.6. Annual precipitation increased from 2780 mm at 4 km to approximately 3500 mm at 13 km from the coast and remained constant through 31 km inland. Precipitation was highest in the late fall and winter months and lowest during the summer. Rates of ion deposition had a similar seasonal pattern to that of precipitation. Concentrations of Cl, SO_4, Mg, Na, and excess Ca (Ca in excess of expected sea salt levels) were highest nearest to the coast and were reflected in higher electrical conductivity in precipitation falling closets to the coast. Vol. 24; no. 2; pp. 239-245

Boatman, J.F.; Valin, V.; Gunter, L.; Laulainen, N.; Lee, R.; Luecken, D.; Ray, J.
  Acid precursor concentrations above the northeastern United States during summer

1987: Three case studies
Journal of Geophysical Research; Jul 20 1990. Description: Two aircraft measured the spatial distribution of SO_2, H_2O_2, O_3, and NO_y above the northeast United States during summer. The aircraft flew three missions between Columbus, Ohio and Saranac Lake, New York. The sulfur dioxide concentration averaged 1-2 ppbv within the planetary boundary layer (pbl) and <1 ppbv in the free troposphere. Superimposed on this background concentration were surface source regions with elevated concentrations as high as 14 ppbv. The hydrogen peroxide concentration averaged 2-4 ppbv, with the highest values found above the top of the planetary boundary layer. The H_2O_2 pattern increased in both size and intensity above the top of the pbl during the afternoon. The ozone concentration varied between 30 and 110 ppbv. The concentration of O_3 increased by as much as 30 ppbv during the day. Nitrogen oxide concentrations were in the range <1-8 ppbv. The pattern of SO_2 and NO_y concentrations remained substantially the same during the daytime hours. Vol. 958; pp. 11831-11845

Boettcher, J; Lauer, S; Strebel, O; Puhlmann, M. Spatial variability of canopy throughfall and groundwater sulfate concentrations under a pine stand. Journal of Environmental Quality; Mar 31 1997 Description: Calculation of solute flux into groundwater by recharge must account for the spatial variability of water flux and recharge solute concentration. Solute concentration of the recharge can be characterized by sampling the uppermost groundwater. In an earlier study under pine (Pinus silvestris L.) on sandy soil, solute concentrations at the water table along a transect fluctuated in a recurrent pattern, probably caused by heterogenous solute input via canopy throughfall. To obtain more insight into the spatial variability of deposition processes and solute concentrations in recharge, transect sampling of the uppermost groundwater was repeated in 1989, and again in 1993. Vol. 26; no. 2; pp. 503-510

Bohonak, A.J.; Wissinger, S.A.
The effects of pond permanence and acidification on the distribution of planktonic invertebrates in a series of high-altitude ponds in central Colorado
Bulletin of the Ecological Society of America; Jun 30 1993
Description: We are studying 20 closely spaced subalpine ponds in Colorado with distinct invertebrate communities. Since 1989, we have monitored distributions and abundances of 54 benthic and 15 planktonic invertebrates, and quantified abiotic factors (water chemistry, pH, temperature, hydroperiod, morphometry) that may explain community differences. We are particularly interested in pH due to a decrease in precipitation pH and an order-of-magnitude range of acid neutralizing capacity at the site.
Vol. 742; pp. 167

Bourdon, J.C.
Improve operations and enhance refinery sulfur recovery.Hydrocarbon Processing; Apr 30 1997. Description: Sulfur is a common contaminant in fossil fuels, released when these fuels are combusted. It causes acid rain and other environmental problems. Sulfur emissions have gained worldwide attention, resulting in tighter requirements for sulfur recovery facilities. New technologies and enhancements to existing technologies have emerged as a result. This overview presents many technologies used for sulfur recovery. It is organized around the unit operations of gas and liquid sweetening, sour water stripping, sulfur recovery, sulfur degassing and solidification, tail gas treating, and incineration. New technical and equipment innovations have resulted in sulfur recovery facilities that are more reliable, recover more sulfur, are easier to operate, and reduce capital and operating costs.
Vol. 76; no. 4; pp. 57-62

Boutacoff, D.
Working with the watershed
EPRI Journal (Electric Power Research Institute); Jan 31 1990
Description: Adding limestone to acidified lakes has proved an effective and

environmentally sound method of restoring populations of game fish and other aquatic life. Building on successful experience with lake liming, researchers are now going a step further by liming a water-shed - the land surrounding a lake - rather than the lake itself. Applied by helicopter, the limestone dissolves and saturated the ground with calcium ions, which neutralize acidic runoff before it reaches the lake. Vol. 151; pp. 28-33

Bowersox, V.C.; Sisterson, D.L.; Olsen, A.R. Acid rain, a world-wide phenomenon: A perspective from the United States International Journal of Environmental Studies; Jan 31 1990. Description: The observation that precipitation carries particles and gases to the earth's surface has been a recurrent interest of scientists for more than two hundred years. Since the late 1940s, this interest has been piqued by the observation that rainfall is acidic, and that this acidity results largely from the byproducts of combustion. Acid rain occurs over large areas of Europe and North America. Vol. 36; pp. 83-101

Bowes, S.M. III; Frank, R.; Swift, D.L. The head dome: A simplified method for human exposures to inhaled air pollutants American Industrial Hygiene Association Journal; May 31 1990.
Description: Acute controlled exposures of human subjects to air pollutants are customarily carried out with whole-body chambers, masks, or mouthpieces. The use of these methods may be limited by cost or technical considerations. To permit a study involving a highly unstable pollutant, artificial acid fog, administered to subjects during natural breathing, a head-only exposure chamber, called a head dome, was developed. It consists of a transparent cylinder with a neck seal which fits over the subject's head and rests lightly on his shoulders. The head dome does not constrain the upper airways or impede exercise on a bicycle ergometer. Ventilation can be monitored accurately and unobtrusively with a neumotachograph at the exhaust port of the dome. A thermocouple may be used to monitor the onset and persistence of oronasal breathing. For short-term exposures to unstable or reactive pollutants lasting up to several hours, the head dome is an effective alternative to a whole-body chamber and probably superior to a face mask or mouthpiece.
Vol. 515; pp. 257-260

Brandt, C; Eldik, R. van
Transition metal-catalyzed oxidation of sulfur(IV) oxides. Atmospheric-relevant processes and mechanisms. Chemical Reviews; Jan 31 1995. Description: The transition metal-catalyzed oxidation of sulfur(IV) oxides has been known for more than 100 years. There is a significant lack of information on the actual role of the transition metal-catalyzed reactions, and much of the earlier work was performed without a detailed knowledge of the chemical system. For this reason attention is focused on the role of transition metal ions in the oxidation of sulfu(IV) oxides in terms of the coordination chemistry involved, as well as the stability & chemical behavior of various participating species. Vol. 95; no. 1; pp. 119-190

Breathing easier: NRDC's work to clear the air Amicus Journal; Jan 31 1994
Description: This article summarizes work of the National Resources Defense Council in a number of important areas involving air pollution reduction: implementation oversight of the Clean Air Acts; Acid Rain Trading program oversight; particulate emissions; and electric vehicles.
Vol. 16; no. 2; pp. 51-52

Breemen, Nico.van.; Finzi, Adrien.C.; Canham, Charles.D. Canopy tree-soil interactions within temperate forests: effects ofsoil elemental composition and texture on species distributions. Canadian Journal of Forest Research; Jul 01 1997. Description: We compared the distribution of adult trees and relatively stable soil properties as part of a study of feedbacks between canopy tree species and soils. In southern New England, soils under Fraxinus americana L. (FRAM) and Acer rubrum L. (ACRU) had high contents of total CaO and MgO.

Under Quercus rubra L. (QURU) and Fagus grandifolia Ehrh. (FAGR), contents of CaO and MgO were low.
Vol. 27; no. 7; pp. 1110-1116

Britton, K.O.; Berrang, P; Mavity, E.
Effects of pretreatment with simulated acid rain on the severity of dogwood anthracnose., Plant Disease; Jun 30 1996 Description: The effects of simulated acid rain on dogwood anthracnose severity were evaluated in a series of greenhouse and field experiments over a 4-year period. In 1990 and 1991, Cornus florida seedlings received 10 weekly foliar applications of simulated rain adjusted to pH 2.5, 3.5, 4.5, and 5.5. They were then placed under mature dogwoods naturally infected with Discula destructive. In both years, the percent leaf area infected increased significantly as the pH of the simulated rain solution decreased.
Vol. 80; no. 6; pp. 646-649

Brook, J.R; Samson, P.J; Sillman, S.
Aggregation of selected three-day periods to estimate annual and seasonal wet deposition totals for sulfate, nitrate, and acidity. Part 1: A synoptic and chemical climatology for eastern North America Journal of Applied Meteorology; Feb 28 1995. Description: Running 3-day periods from 1979 to 1985 were categorized into one of 20 meteorological categories. These categories were developed through the cluster analysis of 3-day progressions of 85-kPa wind flow over eastern North America. The purpose for developing the categories was to identify recurring atmospheric transport patterns that were associated with differing amounts of wet sulfate (SO4(2-)) and nitrate (NO3(-)) deposition at a variety of locations in eastern North America.
Vol. 34; no. 2; pp. 297-325

Brook, J.R; Samson, P.J; Sillman, S.
Aggregation of selected three-day periods to estimate annual and seasonal wet deposition totals for sulfate, nitrate, and acidity. Part 2: Selection of events, deposition totals, and source-receptor relationships. Journal of Applied Meteorology; Feb 28 1995. Description: A method for deriving estimates of long-term acidic deposition over eastern North America based on a limited number of Regional Acid Deposition Model runs has been developed. The main components of this method are the identification of a representative sample of events for model simulation and the aggregation of the deposition totals associated with the events. Meteorological categories, defined according to 3-day progressions of 850-mb wind flow over eastern North America, were used to guide the selection of events.
Vol. 34; no. 2; pp. 326-339

Brooks, R.T.
A regional-scale survey and analysis of forest growth and mortality as affected by site and stand factors and acidic deposition Forest Science; Aug 31 1994
Description: Regression analyses were used to identify factors most closely related to species growth and mortality on continuous forest survey plots in Pennsylvania. In 1985, 200 plots with two prior measurements (in the 1960s and 1970s) were selected and measured for a third time to determine periodic forest growth and mortality rates. Growth and mortality were analyzed for temporal change and for relationship to site, stand, defoliation, and climatic factors and to wet atmospheric deposition.
Vol. 403; pp. 543-557

Brotons, Lluís.; Magrans, Marc.; Ferrús, Lluís.; Nadal, Jacint. Direct and indirect effects of pollution on the foraging behaviour of forest passerines during the breeding season. Canadian Journal of Zoology; Mar 01 1998. Description: Direct and indirect effects of acid deposition on the foraging behaviour of three forest passerine species, the Crested Tit (*Parus cristatus*), Long-tailed Tit (*Aegithalos caudatus*), and Coal Tit (*P. ater*), during the breeding season were studied. In two areas, one affected by pollution from a nearby coal-fired power station and the other unaffected by pollution, we measured needle density and arthropod availability on tree branches; both factors are recognised as affecting the foraging behaviour of forest birds. Tree-site use and movement patterns of birds searching for prey were also determined in both areas. Vol. 76; no. 3; pp. 556-565

Brown, G.W.; Roderick, D.; Nastri, A.
Dry scrubber reduces SO_2 in calciner flue gas. Oil and Gas Journal; Feb 18 1991 Description: This paper discusses the installation of a dry sulfur dioxide scrubber for an existing petroleum coke calciner at its Fruita, Colo., refinery. The dry scrubbing process was developed by the power industry to help cope with the acid rain problem. It is the first application of the process in an oil refinery. The process could also remove SO_2 from the flue gas of a fluid catalytic cracker, fluid coker, or other refinery sources. Vol. 89; pp. 41-44

Brune, W.H; Stevens, P.S; Mather, J.H.
Measuring OH and HO_2 in the troposphere by laser-induced fluorescence at low pressure. Journal of the Atmospheric Sciences; Oct 01 1995. Description: The hydroxyl radical OH oxidizes many trace gases in the atmosphere. It initiates and then participates in chemical reactions that lead to such phenomena as photochemical smog, acid rain, and stratospheric ozone depletion. Because OH is so reactive, its volume mixing ratio is less than 1 part per trillion volume (pptv) throughout the troposphere. Its close chemical cousin, the hydroperoxyl radical HO_2, participates in many reactions as well.
Vol. 52; no. 19; pp. 3328-3336

Bube, R.H.
Materials for photovoltaics
Annual Review of Materials Science; Jan 31 1990. Description: Despite the fact that the 1980s have been a period in which public and government support for photovoltaics has been greatly underemphasized, it is heartening to realize that considerable activity in research and development of solar cells continues, as does a steady, positive experience in the field. In order to focus on the most recent developments in the utilization of materials in photovoltaic devices, this paper is limited primarily to publications from 1987 through 1989. The development of materials for photovoltaic energy conversion is a significant contribution to the development of alternative energy sources. Vol. 20; pp. 19-50

Buhr, M.P.; Parrish, D.D.; Norton, R.B.; Fehsenfeld, F.C.; Sievers, R.E.; Roberts, J.M.
Contribution of organic nitrates to the total reactive nitrogen budget at a rural eastern U.S. site. Journal of Geophysical Research; 6/20/90. Description: Measurements of ambient levels of PAN (peroxyacetyl nitrate or, more properly, peroxyacetic nitric anhydride), C_2-C_5 alkyl nitrates, NO_x (NO + NO_2), HNO_3, and NO_3^- particulates were made concurrently with a measurement of total reactive nitrogen species (NO_y) at Scotia, Pennsylvania, from July 16 to August 31, 1988. Each of the organic nitrates exhibited a marked diurnal variation. The concentration of PAN varied from 0.05 to 3 ppbv (parts per billion by volume). The sum of the alkyl nitrate concentrations varied from less than 2 to 200 pptv (parts per trillion by volume). The NO_y levels varied from 1 to 30 ppbv. During periods of high photochemical activity the contributions to NO_y were 30-45% from HNO_3, 20-25% from PAN, 15-25% from NO_x, 5% from NO_3^-, and an additional 1.5% from the C_2-C_5 alkyl nitrates, on average.
Vol. 957; pp. 9809-9816

Bukaveckas, Paul.; Shaw, William.
Phytoplankton responses to nutrient and grazer manipulations among northeastern lakes of varying pH.
Canadian Journal of Fisheries and Aquatic Sciences; Apr 01 1998.
Description: Short-term nutrient enrichment and zooplankton exclosure experiments were conducted at 14 lakes representing various stages of acidification (pH 4.6-6.8). We measured changes in chlorophyll as an indicator of the severity of nutrient limitation and grazing intensity and compared these with independent measures of P limitation (cell P quotas and phosphatase activity) and grazing (zooplankton densities and inferred

community grazing rates). Results from nutrient enrichment experiments showed good correspondence to measured phosphatase activity but not cell P quotas.
Vol. 55; no. 4; pp. 958-966

Bulger, Arthur.J.; Cosby, Bernard.J.; Webb, James.R. Current, reconstructed past, and projected future status of brook trout (*Salvelinus fontinalis*) streams in Virginia. Canadian Journal of Fisheries and Aquatic Sciences; Jul 01 2000.
Description: Southern Appalachian streams host a rich diversity of fishes, but the Southern Appalachian Assessment concluded that 70% of stream locations showed significant fish community degradation, partly due to acid deposition. About 40% of total Southern Appalachian trout stream length occurs in Virginia. Our research in Shenandoah National Park, Virginia, has documented both chronic and episodic acidification in streams and brook trout (*Salvelinus fontinalis*) mortality during acid episodes.
Vol. 57; no. 7; pp. 1515-1523

Burger, J; Cooper, K; Martin, M.
Attitudes toward environmental hazards: Where do toxic wastes fit?
Journal of Toxicology and Environmental Health; Jan 31 1997
Description: The public is continually faced with making decisions about the risks associated with environmental hazards, and, along with managers and government officials, must make informed decisions concerning possible regulation, mitigation, and restoration of degraded sites or other environmental threats. We explored the attitudes regarding several environmental hazards of six groups of people: undergraduate science majors, undergraduate nonscience majors, and graduate students in environmental health, in ecological risk assessment, and in nonscience disciplines, as well as nonstudents over 35 yr of age. We had predicted that there would be significant differences in attitudes between science and nonscience majors and as a function of age. Vol. 51; no. 2; pp. 109-121

Burkhardt, D.A.
Acid rain: What the final bill looks like
Public Utilities Fortnightly; Dec 06 1990
Description: This article examines the possible financial impacts of the 1990 amendments to the Clean Air Act. Topics include an overview of the bill, impact on utility companies, and implications for utility investors. The author feels the bill has important implications for investors who own utility stocks, particularly the stocks of coal-burning utilities in the Midwest and Appalachia.
Vol. 12612; pp. 56-57

Burkhart, L.A.
Action and auction: The SO_2 emissions allowances. Public Utilities Fortnightly; Jun 15 1993. Description: On January 11, the Environmental Protection Agency (EPA) published its final comprehensive rule on acid rain in the Federal Register, directing that sulfur dioxide (SO_2) emissions reductions are to occur in two phases. On March 5, the EPA issued its final rule allocating emissions allowances for use after 2000, a rule intended to facilitate pollution reduction by putting into effect the emissions allowance trading system created by the Clean Air act Amendments of 1990 (CAAA). The flurry of activity that followed these rules is not surprising, although the direction of some of that activity is. Vol. 13112; pp. 42-44

Burkhart, L.A.
EPA allocates emission allowances for phase II plants
Public Utilities Fortnightly; Apr 15 1993
Description: The Environmental Protection Agency (EPA) last month issued a final rule allocating acid rain emission allowances for use after 2000 by most United States power plants. The rule sets the stage for significant pollution reduction through the emission trading system established by the Clean Air Act Amendments of 1990. The rule complements EPA's comprehensive, final acid rain rule for SO_2, which was published in the January 11, 1993, Federal Register.
Vol. 1318; pp. 53

Burton, A.J.; Pregitzer, K.S.; Reed, D.D.
Leaf area and foliar biomass relationships in northern hardwood forests located along an 800 km acid deposition gradient
Forest Science; Sep 30 1991
Description: The canopies of northern hardwood forests dominated by sugar maple (Acer saccharum Marsh.) were examined at five locations spanning 800 km along an acid deposition and climatic gradient in the Great Lakes region. Leaf Area Index (LAI) calculated from litterfall ranged from 6.0 to 8.0 in 1988, from 4.9 to 7.9 in 1989, and from 5.3 to 7.8 in 1990. The data suggest that maximum LAI for the sites is between 7 and 8. Insect defoliation and the allocation of assimilates to reproductive parts in large seed years reduced LAI by up to 34%.
Vol. 374; pp. 1041-1059

CAA update: New HAP, NO_x rules: more to come. Hazmat World; Apr 30 1994
Description: In March of 1994 EPA announced new air toxics rules for chemical companies, electrical utilities and several manufacturing sectors. In one of five major rulings issued under authority of the 1990 CAA Amendments, the Agency gave chemical producers three years to achieve an 88 percent reduction (compared to 1990 levels) in air toxics, or hazardous air pollutant (HAP) emissions, at their facilities. The ruling is expected to affect about 370 chemical plants in 38 states. EPA estimates that total capital costs to the chemical industry for compliance with the rule will be about $450 million, with annual costs of $230 million.
Vol. 74; pp. 10-11

Cahill, T.A.
Combined atomic-nuclear-optical-mass techniques for aerosol analysis
Transactions of the American Nuclear Society; Nov 30 1991
Description: Anthropogenic particulate matter in ambient atmosphere, even at locations well removed from urban and industrial areas, is often the dominant factor in visibility reduction, toxic metal transport, and acid-rain precursors. The full compositional analysis of such aerosols is vital to establishing aerosol impacts. However, the task is difficult because of the small amount of fine aerosol mass (a few micrograms per cubic meter of air), the wide range of elements needed for characterization of the gravimetric mass, and the extremely low levels of tracer elements. At the University of California, Davis, the authors have emphasized full analysis of aerosol mass at up to 105 remote sites in the US as part of the interagency monitoring of protected visual environments (IMPROVE) fine aerosol network, achieving an average of 93% of the mass reconstructed from atomic and nuclear ion beam techniques. IMPROVE is a joint effort of the National Park Service, the Environmental Protection Agency, the Forest Service, the Bureau of Land Management, and the Fish & Wildlife Service. Vol. 64; pp. 209-210

Campbell, D.H.; Turk, J.T.; Spahr, N.E.
Response of Ned Wilson Lake watershed, Colorado, to changes in atmospheric deposition of sulfate
Water Resources Research; Aug 31 1991
Description: The Ned Wilson Lake watershed responds directly and rapidly to changes in precipitation inputs of sulfate, which has important implications for effect of acid deposition on the aquatic system. Chemistry at three precipitation collection sites and three watershed sites (a pond, a lake, and a spring) has been monitored in and near the Flattops Wilderness Area in northwestern Colorado beginning in 1981-1983. Vol. 278; pp. 2047-2060

Cannon, W.N. Jr.
Gypsy moth (Lepidoptera: Lymantriidae) consumption and utilization of northern red oak and white oak foliage exposed to simulated acid rain and ozone
Environmental Entomology; Jun 30 1993
Description: Two-year-old seedlings of white oak, Quercus alba L., and red oak, Q. rubra L., were exposed to ozone (O_3) fumigations in four continuously stirred tank reactor chambers in the greenhouse for 8 h/d, 3 d/wk for 6 wk. Fumigation treatments were charcoal-filtered air (CFA) and CFA + 0.15 ppm O_3. Two simulated

rain treatments, pH 4.2 and pH 3.0, of-1.25 cm were applied once each week in rain-simulation chambers. Gypsy moth, Lymantria dispar (L.), third instars were allowed to feed on leaf disks from treated seedlings for 24 h. Leaf area consumed, food assimilated, weight gain, and relative growth rate (RGR) were examined.
Vol. 223; pp. 669-673

Carignan, Richard.; Steedman, Robert.J.
Impacts of major watershed perturbations on aquatic ecosystems
Canadian Journal of Fisheries and Aquatic Sciences; Feb 01 2000
Description: This Supplement presents data syntheses and new evidence from temperate (primarily boreal) North American studies of aquatic ecosystem response to episodic watershed deforestation and acid rain. These studies confirm the dominant role of the watershed in modulating aquatic response to terrestrial disturbance and quantify important regional differences related to physiography, vegetation, and drainage patterns. Comparisons of watershed disturbance by wildfire and logging revealed both similarities and differences in aquatic impact and underscore the need for ongoing regional evaluation of forest management models based on simulation of natural disturbance patterns.
Vol. 57; no. 2; pp. 1-4

Cascio, T.B.
Fuel type as a factor influencing compliance by utilities under the Acid Rain Programme. Journal of the Air and Waste Management Association; May 01 1999
Description: Passage of the 1990 Clean Air Act Amendments launched the Acid Rain Program in the United States. This initiative, based on the market mechanism of a sulfur dioxide tradable allowance system, was a dramatic departure from traditional command and control strategies designed to reduce air pollution emissions. Power plant managers have flexibility under the program to select and implement a variety of options to reduce emissions below mandated levels.
Vol. 49; no. 5; pp. pp.562-568

Cason, T.N.; Plott, C.R.
EPA's new emissions trading mechanism: A laboratory evaluation
Journal of Environmental Economics and Management; Mar 31 1996
Description: The centerpiece of the acid rain control program in the Clean Air Act Amendments of 1990 is a system of tradable emission permits. Utilities must hold permits to emit sulfur dioxide, and the number of available permits will decline over time to reduce total emissions. This paper reports 12 laboratory markets that investigate trader behavior in this new institution and evaluate its performance relative to the more commonly observed uniform price call mmarket The uniform price version is found to be more efficient, induces more truthful revelation of underlying values and costs, provides more accurate price information, and is more responsive to and recovers more quickly from changes in the underlying market conditions. Vol. 30; no. 2; pp. 133-160

Castelle, A.J.; Galloway, J.N.
Carbon dioxide dynamics in acid forest soils in Shenandoah National Park, Virginia. Soil Science Society of America Journal; Jan 31 1990. Description: Carbonic acid, derived from soil CO_2, may strongly affect streamwater chemistry. The spatial and temporal variability of soil CO_2 levels was examined in the White Oak Run watershed of Shenandoah National Park, Virginia. This watershed is in a mountainous, forested region that is being studied to monitor the effects of atmospheric acidic deposition.
Vol. 541; pp. 252-257

Caton, J.E.; Ho, C.H.; Williams, R.T.; Griest, W.H. Characterization of insoluble fractions of TNT transformed by composting. Journal of Environmental Science and Health, Part A: Environmental Science and Engineering; May 31 1994
Description: Soil contaminated with explosives was supplemented with carbon-14 labelled 2,4,6- trinitrotoluene (^{14}C-TNT) and was composted in a field static pile composting experiment. After 90 d of composting, the distribution of carbon-14

(^{14}C) activity in fractions from acetonitrile extraction ("free" fraction, 1.2% of the initial ^{14}C-activity) and filtration ("insoluble-particle" fraction, 17.9%), alkaline hydrolysis ("insoluble-hydrolyzable" fraction, 56.8%), and combustion of the residue ("insoluble-nonhydrolyzable" fraction, 4.7%) showed that the bulk of the ^{14}C-activity, and presumably transformed product(s) of the ^{14}C-TNT, accumulated in a nonextractable, but hydrolyzable fraction.
Vol. 294; pp. 659-670

Chapman, E.G.; Luecken, D.J.
Associations between pollutant emissions and precipitation chemistry: An empirical analysis. Journal of Geophysical Research; Jan 20 1993. Description: Empirical analyses were conducted of the associations between anthropogenic SO_2 emissions and precipitation SO_4^{2-} concentrations as well as between anthropogenic NO_x emissions and precipitation NO_3^- concentrations across central North America. Both the 1982 Electric Power Research Institute (EPRI) National Air Pollutant Emissions Inventory and the 1985 National Acid Precipitation Assessment Program (NAPAP) Emissions Inventory, along with precipitation chemistry data from up to 130 monitoring sites, were used in the analyses.
Vol. 981; pp. 1113-1122

Cheplick, G.P.
Effect of simulated acid rain on the mutualism between tall fescue (Festuca arundinacea) and an endophytic fungus (Acremonium coenophialum)
International Journal of Plant Sciences; Mar 31 1993. Description: Biotic interactions between plants and microorganisms have the potential to be affected by acidic precipitation. I examined the effect of simulated sulfuric acid rain on the mutualism between a perennial forage grass (Festuca arundinacea) and a fungal endophyte (Acremonium coenophialum). Acid water was supplied as mists sprayed onto leaf surfaces or as water added to the soil for two groups in a greenhouse: one group had high levels of endophyte infection, while the other was predominantly noninfected. Control plants received distilled water (pH 6), while others received sulfuric acid water at pH 4.5 or pH 3. Vol. 1; pp. 134-143

Chesnoy, A.B.; Pack, D.J.
S_8 threatens natural gas operations, environment. Oil and Gas Journal; Apr 28 1997. Description: The presence of elemental sulfur (S_8) in natural-gas can have serious consequences for gas production, processing, and pipeline operations. Yet, no reliable means exist to analyze a gas stream for its presence. Elemental sulfur within a gas stream can be costly not only because of equipment damage and resulting poor reliability, but also because of extensive plant downtime. S_8 has been found to affect the accuracy of flow-measuring instruments, to damage turbine blades because of passage of loose S_8 particles, and to cause severe plugging of exchangers. Vol. 95; no. 17; pp. 74-79

Chowdhury, B.H.
Emission control alternatives for electric utility power plants
Energy Sources; Jun 30 1996
Description: Recently, the electric power industry has come under close environmental scrutiny. It is estimated that more than half of the sulfur dioxide (SO_2) emissions and a third of the nitrogen oxide (NO_x) emissions in the US are by-products of fossil-fuel-fired generating stations. Future estimates predict an even higher percentage of use of coal, the principal pollutant among the fossil fuels, in world electricity production, particularly in the industrially developed countries of the world. There is little doubt that electric utilities will have to become more involved in finding ways for abatement of these precursors for acid rain.
Vol. 18; no. 4; pp. 393-406

Clair, T.A.; Pollock, T.L.; Collins, P.V.; Kramer, J.R.
How brown waters are influenced by acidification: The HUMEX lake case study
Environment International; Jan 31 1992
Description: In order to assess how brown

waters will react with changes in acid inputs caused by acid precipitation, it is important to estimate quantitatively the concentration of organic anions in solution. A new method of determining the organic anion component of brown waters was developed. The approach was then tested on water samples collected in Nova Scotia. Vol. 186; pp. 589-596

Clark, E.
Forest Industry: Timberlands tomorrow American Forests; Jan 31 1990 Description: This article describes the trend towards environmentalism as it affects industrial forestry. Foresters will need to utilize techniques in genetic engineering to increase productivity; they will also need to develop species that can adapt to global warming and be more resistant to air pollutants and acid rain. At present, whole-tree logging practices are taking the place of slash burning and forest residuals are being used as boiler fuel. Vol. 9656; pp. 58-60, 68-71

Clark, P.A.; Gervat, G.P.; Hill, T.A.; Marsh, A.R.W.; Chandler, A.S.; Choularton, T.W.; Gay, M.J.
A field study of the oxidation of SO_2 in cloud. Journal of Geophysical Research; Aug 20 1990. Description: This paper describes the execution and analysis of one of a series of field experiments performed at Great Dun Fell in the northern Pennines of England designed to study the oxidation of SO_2 in cloud. The experiments are designed to isolate the effect on cloud chemistry by artificially releasing SO_2 gas into the air entering cloud base. Observations made in cloud on April 30, 1986, clearly and unequivocally demonstrated the production of sulfate ions in cloudwater. Vol. 959; pp. 13,985-13,995

Clean-burning fuels produced from low-grade coal
Air Pollution Consultant; Jan 31 1995 Description: Under the acid rain program, operators of large combustion units are required to reduce their emissions of sulfur oxides (SO_x) and nitrogen oxides (NO_x). Although the program provides significant flexibility through its system of marketable emission allowances, regulated sources needing to reduce SO_x emissions typically choose one of the following two options: (1) switch to a low- sulfur fuel, or (2) add end-of-pipe controls. Because it is naturally low in sulfur, one good candidate for fuel switching is coal mined from the Powder River basin in northeast Wyoming. The Encoal Corporation (Gillette, Wyoming) has attempted to improve the economics of using Powder River coal by installing a coal liquification plant at an existing mine near Gillette. The plant, cofunded by the Department of Energy (DOE) and Zeigler Coal Holding Company (the parent company to Encoal), has demonstrated commercial-scale application of a liquids- from-coal (LFC) process developed by SGI International. The LFC process represents a middle-of-the-road approach to coal treatment. As described, here, the process converts high-moisture, low-grade coal into process-derived fuel (PDF- an upgraded solid coal product) and coal-derived liquids (CDL-fuel-oil type liquids). The LFC process also produces an organic gas stream, which is burned internally as an energy source. Finally, the LFC process can be adapted, if necessary, to remove sulfur from high-sulfur coal. Vol. 52; pp. 1.6-1.8

Clean-burning fuels produced from low-grade coal
Air Pollution Consultant; Mar 31 1995 Description: Under the acid rain program, operators of large combustion units are required to reduce their emissions of sulfur oxides (SO_x) and nitrogen oxides (NO_x). Although the program provides significant flexibility through its system of marketable emission allowances, regulated sources needing to reduce SO_x emissions typically choose one of the following two options: (1) switch to a low-sulfur fuel, or (2) add end-of-pipe controls. Because it is naturally low in sulfur, one good candidate for fuel switching is coal mined from the Powder River basin in northeast Wyoming. Vol. 5; no. 2; pp. 1.6-1.8

Clinton, P.W.; Allen, R.B.; Davis, M.R.
Nitrogen storage and availability during stand development in a New Zealand *Nothofagus* forest
Canadian Journal of Forest Research; Feb 01 2002. Description: Stemwood production, N pools, and N availability were determined in even-aged (10, 25, 120, and >150-year-old) stands of a monospecific mountain beech (*Nothofagus solandri* var. *cliffortioides* (Hook. f.) Poole) forest in New Zealand recovering from catastrophic canopy disturbance brought about by windthrow. Nitrogen was redistributed among stemwood biomass, coarse woody debris (CWD), the forest floor, and mineral soil following disturbance. Vol. 32; no. 2; pp. 344-352

Coal preparation
Mining Engineering (Littleton, Colorado); May 31 1991
Description: The acid rain control legislation has prompted the Department of Energy (DOE) to seek new technology using the Clean Coal Technology program solicitation. The main goal of the program is to reduce SO_2 emissions below 9 Mt/a (10 million stpy) and NO_x emission below 5.4 Mt/a (6 million stpy) by the year 2000. This would be accomplished by using precombustion, combustion, post combustion and conversion technology. Vol. 435; pp. 523-524

Coal preparation: The foundation for modern coal use
PETC Review (Pittsburgh Energy Technology Center); Jan 31 1992
Description: In the US Department of Energy's (DOE's) coal research and development (R and D) program, coal preparation holds a unique position. Coal preparation is the enabling technology, serving as the foundation for coal use across the full spectrum of current and future applications of coal. This article introduces DOE's Coal Preparation Program. Emphasis is on the key considerations that drive public investment in research and technology development directed at advanced processes for cleaning coal. The background for this discussion is the history of coal preparation technologies, the current state of the industry, and the impact of the Clean Air Act Amendments of 1990. Vol. 5; pp. 4-13

Coichev, N.; R., Van Eldik,
Kinetics and mechanism of the sulfite-induced autoxidation of cobalt(II) in aqueous azide medium
Inorganic Chemistry; May 15 1991
Description: The autoxidation of Co(II) in azide medium is accelerated by sulfur(IV) oxides at a concentration level of 10^{-5} M. The formation of Co(III) was followed spectrophotometrically under the following conditions: $(Co(II)) = 5 \times 10^{-4}$ M; initial $(Co(III)) = (0-3.6) \times 10^{-5}$ M; (total $S(IV)$) = $(1-4) \times 10^{-5}$ M; $4 < pH < 6$; (total N_3^-) = 0.1 - 0.5 M; temperature = 25C; ionic strength = 1.0 M. The autoxidation reaction exhibits typical autocatalytic behavior in which the induction period depends on the Co(III) concentration. Vol. 3010; pp. 2375-2380

Cole, F.
Environmental consequences of increased natural-gas usage
United States Geological Survey, Professional Paper; Jan 31 1993
Description: Energy use is the primary cause of many environmental problems in the United States and around the world. Fossil fuels, including coal, oil, and natural gas, supply roughly 90 percent of our energy needs in the United States, and they are directly responsible for urban and industrial air pollution and acid rain. Combustion emissions from fossil fuels also contribute to the Earth's greenhouse effect, and they may play an important role in ozone depletion in the stratosphere, and oxidant depletion in the troposphere. Vol. 1570; pp. 619-634

Coleman, W.G.
Biodiversity and industry ecosystem management. Environmental Management; Nov 30 1996. Description: Biodiversity describes the array of interacting, genetically distinct populations and species in a region, the communities they are functioning parts. Ecosystem health is a process identifying biological indicators,

end points, and values. The decline of populations or species, an accelerating trend worldwide, can lead to simplification of ecosystem processes, thus threatening the stability an sustainability of ecosystem services directly relevant to human welfare in the chain of economic and ecological relationships. The challenge of addressing issues of such enormous scope and complexity has highlighted the limitations of ecology-as-science.
Vol. 20; no. 6; pp. 815-825

Collins, M.; Terhune, K.
A model solution for tracking pollution Chemical Engineering (New York); Jun 30 1994. Description: In the US, operators throughout the chemical process industries are being pressed by a strong environmental force--the 1990 Clean Air Act Amendments (CAAA). Far broader than the narrowly focused Clean Air Act of 1970, the latest amendments have been designed to sharply reduce smog, acid rain and various environmental pollutants in the earth's atmosphere. Combined with individual regional and state regulations, the CAAA declare that all major sources of air pollutants (plants emitting 10 tons/yr of any listed pollutant, or 25 tons/yr overall) will be compared against 12% of their competitors' most tightly controlled plants and will be expected to perform at that benchmark level. Vol. 101; pp. 32-37

Collins, P.V.; Kramer, J.R.; Collins, D.J.; Sayer, B.G.
Soil-water interactions at the HUMEX Lake Skjervatjern. Environment International; Jan 31 1992. Description: Changes in the nature of humic substances at an experimentally acidified lake are determined from titration data reduction for functional group analysis. The acidified portion of the catchment has increased in carboxyl functional groups (27 meq/g C) compared to the control ([approximately]10 meq/g C). The organic rich peat on land has been depleted of acidic functional groups and has markedly increased in phenolic sites compared to the control portion of the catchment.
Vol. 186; pp. 565-576

Conkling, B.L.; Blanchar, R.W.; Niblack, T.L.
Effects of foliar and soil acidity on the rhizospere pH of alfalfa, corn and soybean Journal of Environmental Quality; Jan 31 1991. Description: The acidity of ambient rainfall and its effect on soil and plants is a growing concern. Glass microelectrodes were used to investigate the effect of soil pH and foliar application of acid rain on the rhizosphere pH of alfalfa (Medicago sativa L. cv. Arrow), corn (Zea mays L. cv. B72 x MO17), and soybean (Glycine max (L.) Merr. cv. Williams 82). Plant roots were grown in minirhizotrons containing a reformed sample of Seymour silt loam A horizon over a silty clay loam Bt horizon. Low and high pH levels of 4.9 and 6.2 in the A horizon and 4.0 and 5.7 in the Bt horizon were established using dilute sulfuric acid or calcium ozide, respectively. Plants received daily applications of simulated rain, which was either acid (pH 3.1) or non-acid (pH 5.6). After 5,6, or 15 d of foliar applications to corn, soybean, or alfalfa, respectively, the rhizosphere pH was measured using a glass microelectrode. The pH values for corn and soybean increased with distance from the root while the pH values for alfalfa decreased with distance. As the soil pH increased from 4 to near 6, the difference between the pH at the root surface and the bulk soil increased from 0 to near 1.
Vol. 202; pp. 381-386

Connick, R.E.; Lee, Shaoyung; Adamic, R.
Kinetics and mechanism of the oxidation of HSO_3^- by HSO_5^-
Inorganic Chemistry; Mar 03 1993 Description: The oxidation of bisulfite ion is of importance in flue gas desulfurization processes and in the conversion of gaseous SO_2 into acid rain in water droplets in the atmosphere. The oxidation of bisulfite ion by peroxymonosulfate ion, HSO_5^-, to form sulfate ion has been studied in the pH region 3.8-7.9. Vol. 325; pp. 565-571

Corcoran, E.
Cleaning up coal
Scientific American; May 31 1991 Description: According to the percentages, coal is still King. Coal-fired power plants

generate more than 50% of US electricity. But every year those utilities also pour forth 70% of the sulfur dioxide and significant portions of other pollutants that cause acid rain and contribute to global warming. Now the US is trying a novel market-based approach to reducing those emissions. The policy of granting tradable allowances for polluting is driving electric utilities to consider a whole range of technologies for reducing emissions. The paper describes the policy, some of the clean coal technologies being considered, and assesses whether the policy will work. Vol. 2645; pp. 106-116

Cozza, A.; Faulkner, K.F.
Acid rain program offers free-market incentives, portends future regulation Hazmat World; May 31 1993
Description: The burning of fossil fuels, particularly coal and oil, results in emissions of sulfur dioxide (SO_2) and nitrogen oxide (NO_x). Such emissions and their by-products damage ecosystems and man-made materials, and threaten human health. In 1985, 23 million tons of SO_2 and 19 million tons of NO_x were emitted from US sources. Nearly 70% of the SO_2 and almost 40% of the NO_x emissions originated from electric utilities. Title IV of the CAA Amendments requires EPA to establish an acid rain program designed to reduce SO_2 and NO_x emissions from electric utility plants. Vol. 65; pp. 37-4

Cronan, C.S.; Driscoll, C.T.; Newton, R.M.; Kelly, J.M.; Schofield, C.L.; Bartlett, R.J.; April, R. A comparative analysis of aluminum biogeochemistry in a northeastern and a southeastern forested watershed. Water Resources Research; Jul 31 1990. Description: This comparative biogeochemical analysis focused on the patterns and processees of aluminum cycling in two small watersheds, one in the west-central Adirondacks of New York and the other on Cumberland Plateau of eastern Tennessee. Despite shared similarities in soil acidity, soil exchangeable aluminum concentrations, and elevated inputs of acidic deposition, the northern and southern sites exhibited strong differences in aqueous aluminum chemistry and transport. Soil and stream drainage waters in the northern watershed were more acidic, and contained higher concentrations of base cations, sulfate, nitrate, and organic carbon than waters in the southern ecosystem. Mean concentrations of biologically active labile inorganic aluminum, Al_i, ranged from 17 to 46 $\{mu\}mol\ L^{-1}$ in soil solutions and stream water in the northern drainage basin, and from 0 to 2 $\{mu\}mol\ L^{-1}$ in the southern system. The major differences in aluminum chemistry and transport between the two watersheds were related to different patterns of alkalinity generation and mobile anion transport in these contrasting systems. In the northern watershed, atmospheric inputs of acidicty were partially neutralized through the release of mixed cations from soils and detritus. Because of the high mobility of sulfate and nitrate in the northern watershed, there was significant transport of Al through the soil profile and into stream water. At the southern watershed, soil sulfate adsorption, biological retention of nitrate, and base cation release were the major sources of acid neutralizing capacity for soil drainage waters and surface waters. Vol. 267; pp. 1413-1430

Cronan, C.S.; Schofield, C.L.
Relationships between aqueous aluminum and acidic deposition in forested watersheds of North America and northern Europe. Environmental Science and Technology; Jul 31 1990. Description: There are important interregional differences in the concentrations of aqueous labile aluminum in soil waters and surface waters of forested watersheds in eastern North America and northern Europe. These variations in labile aluminum are a function of several interacting factors, including soil base saturation, solution pH, anion chemistry, pH-dependent solubility and adsorption relationships, and soil hydrology. The highest concentrations of labile aluminum are found in watersheds characterized by soils with <10-15% base saturation and elevated concentrations of strong acid

anions. For such sites, there is a strong positive relationship between acidic deposition, as a source of mobile anions, and aluminum mobilization. Future increases or decreases in mobile strong acid anions at these types of watersheds are likely to produce corresponding changes in both the concentrations of labile aluminum and the potential for aluminum toxicity to organisms. Vol. 247; pp. 1100-1105

Cunjak, R.A.; Prowse, T.D.; Parrish, D.L.
Atlantic salmon (*Salmo salar*) in winter: "the season of parr discontent"?
Canadian Journal of Fisheries and Aquatic Sciences; Jan 01 1999
Description: Winter is a dynamic period. Effects of the winter regime on northern streams and rivers is extremely variable and characterized by dramatic alterations in physical habitat to which Atlantic salmon (*Salmo salar*) must acclimate and adapt to survive. In this paper, we synthesize recent advances in the biological and hydrologic/ geomorphic disciplines, with specific reference to Atlantic salmon overwintering in the freshwater portions of those running waters subject to freezing water temperatures. Vol. 56; no. 1; pp. 161-180

Curtin, D; Campbell, C.A; Messer, D.
Prediction of titratable acidity and soil sensitivity to pH change. Journal of Environmental Quality; Nov 30 1996.
Description: The buffering capacity of a soil must be known if we are to model changes in its pH due to fertilization or acidic deposition. Using a diverse suite of 59 agricultural soils from Saskatchewan, Canada, we attempted to develop a quantitative index of soil sensitivity to pH change. We measured pH changes after equilibration with several levels of NH_4OH and estimated titratable acidity to pH 7, 8, and 9. Titratable acidity corresponded to the nonexchangeable component of acidity because, with one exception, no KCl-extractable acidity was detected.
Vol. 25; no. 6; pp. 1280-1284

Czarnecki, J.; Pereira, C.J.; Uberoi, M.; Zak, K.P.

Put a lid on NO_x emissions
Pollution Engineering; Nov 30 1994
Description: The effect of nitrogen oxides (NO_x) emissions on air quality and the subsequent need for NO_x control has been a subject of significant debate over the years. Although recognized as a contributor to both acid rain and ground level ozone formation, NO_x has received only limited regulatory focus in the US. The domestic situation, however, is changing rapidly largely because of the Clean Air Act Amendments (CAAA) of 1990. Vol. 2612; pp. 26-31

Dahlgren, R.A.; McAvoy, D.C.; Driscoll, C.T.
Acidification and recovery of a spodosol Bs horizon from acidic deposition
Environmental Science and Technology; Apr 30 1990. Description: A laboratory study was conducted to examine acidification and recovery of a Spodosol Bs horizon from acidic deposition in the Bear Brook Watershed (BBW) in central Maine. A mechanical vacuum extractor was used to draw solutions through a soil column at three treatments containing 40, 100, or 160 {mu}mol/L SO_4^{2-}. Following 44 days of leaching, all treatments were decreased to the 40 {mu}mol/L SO_4^{2-} level to examine recovery from acidification. Acid additions were initially neutralized by release of basic cations and sulfate adsorption. Following attainment of steady-state conditions for basic cations and SO_4^{2-} with respect to the soil adsorption complex, Al dissolution was the primary neutralization mechanism. Aqueous Al activities appeared to be regulated by equilibrium with an $Al(OH)_3$ mineral phase. Following decreases in acid loadings, recovery was rapid resulting in retention of basic cations, reversible release of SO_4^{2-}, and a marked reduction in the concentrations of soluble Al.
Vol. 244; pp. 531-537

Dahlin, R.S.; Snyder, T.R.; Bush, P.V.
Effects of sorbent injection on particulate properties: Part II. High-temperature sorbent injection. ir and Waste; Jan 31 1993. Description: To comply with the Acid Rain Provisions of the 1990 Clean

Air Act Amendments, many coal-burning utilities are considering the installation of sorbent injection processes. The effects of sorbent injection processes on existing equipment for particulate control must be evaluated. This paper reviews the effects on particulate properties of high-temperature sorbent injection processes. Vol. 43; pp. 91-96

Dai, Q; Freedman, A; Robinson, G.N. Sulfuric acid-induced corrosion of aluminum surfaces. Journal of the Electrochemical Society; Dec 31 1995. Description: The sulfuric acid-induced corrosion of smooth (2 nm average roughness) aluminum surfaces has been studied in real times using an in situ Fourier transform infrared reflection absorption spectrometer and a quartz crystal microbalance. Submicron thick, 35 to 55 weight percent (5 to 12 molal), sulfuric acid films were formed on room temperature metal surfaces by the reaction of gas-phase SO_3 and H_2O vapor in a flowing gas system at a total pressure of ~200 Torr. The deposition of the acid films and subsequent changes in their chemical composition resulting from corrosion of the aluminum substrate could be monitored using characteristic infrared absorption features. Vol. 142; no. 12; pp. 4063-4069

Dalledone, E.; Barbosa, M.A.; Wolynec, S. Zinc-55% aluminum-1.6% silicon coating compared with zinc coating Materials Performance; Jul 31 1995 Description: A comparative investigation of zinc-55% aluminum-1.6% silicon (Zn-55Al-1.6Si) alloy and zinc coatings, both applied by hot-dip process on low-carbon steel, was performed, with special attention to the protection provided to the substrate metal. In all tests, the performance of the alloy-coated steel was superior to that of galvanized steel. The electrochemical tests did show that both coatings provide cathodic protection to the basis metal; the galvanic potentials are equal to -1,050 and -900 mV (saturated calomel electrode) for zinc and the alloy, respectively, which are adequate to keep the steel inside the immunity region. Vol. 34; no. 7; pp. 24-28

Darveau, Marcel.; Martel, Jocelyn.; DesGranges, Jean-Luc.; Mauffette, Yves. Associations between forest decline and bird and insect communities in northern hardwoods. Canadian Journal of Forest Research; Jun 01 1997. Description: Forest decline can be the result of multiple stresses such as air pollution or natural processes. This phenomenon can directly affect trees as well as the fauna and nonarborescent flora within these ecosystems. From 1987 to 1989 we studied insectivorous bird, arthropod, and plant populations in 18 sugar maple (Acer saccharum Marsh.) stands affected by decline in southern Quebec. On average, there was a 20-30% foliage loss. Vol. 27; no. 6; pp. 876-882

David, M.B.; Fasth, W.J.; Vance, G.F. Forest soil response to acid and salt additions of sulfate. II. Aluminum and base cations. Soil Science; Mar 31 1991. Description: Reconstructed Spodosol and intact Alfisol soil columns were used to examine the effects of 52 weeks of additions of various simulated throughfall solutions on base cation, Al, acid neutralizing capacity, and pH levels in soil leachates. The authors purpose was to determine the effects of acid & salt additions of SO_4^{2-} on cation leaching in two forest soils and to investigate the influence of episodic events of seasalt and pCO_2. Vol. 1513; pp. 208-219

David, M.B.; Vance, G.F.; Fuller, R.D.; Fernandez, I.J.; Mitchell, M.J.; Stam, A.C.; Nodvin, S.C.. Spodosol variability and assessment of response to acidic deposition Soil Science Society of America Journal; Jan 31 1990. Description: Variability in forest soils makes it difficult to observe short-term changes in chemical properties under field conditions. A buried soil-bag technique was developed to examine the chemical response of a Maine forest soil to loadings of strong acids (HNO_3 and H_2SO_4). The buried soil-bag technique detected small alterations in forest soil chemistry under field conditions, with minimal disturbance to study plots. Vol. 542; pp. 541-548

Davis, D.D. Skelly, J.M.
> Growth response of four species of Eastern hardwood tree seedlings exposed to ozone, acidic precipitation, and sulfur dioxide. Journal of the Air and Waste Management Association; Mar 31 1992. Description: In 1987 a study was conducted in controlled environment chambers to determine the foliar sensitivity of tree seedlings of eight species to ozone and acidic precipitation, and to determine the influence of leaf position on symptom severity. Jensen and Dochinger conducted concurrent similar studies in Continuously Stirred Tank Reactor (CSTR) chambers with ten species of forest trees. Based on the results of these initial studies, four species representing a range in foliar sensitivity to ozone were chosen: black cherry (Prunus serotina Ehrh.), red maple (Acer rubrum L.), northern red oak (Quercus rubra L.) and yellow-poplar (Liriodendron tulipifera L Vol. 423; pp. 309-311

Davis, P.N.
> Effect of Clean Air Act Amendments of 1990 on use of Midwestern coal. Geological Society of America, Abstracts with Programs; Mar 31 1993. Description: The acid rain provisions of the Clean Air Act Amendments of 1990 (42 U.S.C. [section][section] 7,651--7651o) and implementing regulations of October 1992 will substantially modify use of high-sulfur coal by utilities during the next decade. The Act adopts a market-based approach, allowing utilities to meet those emission levels by (1) installing scrubbers, low-emission boilers, or coal- cleaning technology, (2) switching to lower-sulfur coal, or (3) purchasing emission allowances to cover excess emissions. Those allowances will be sold by utilities which have reduced emissions below required levels. Initial allowances are distributed according to a statutory formula to existing plants based on 1985 outputs and to new plants beginning operation before 2000. Vol. 253; pp. 16

Deckert, Ron.J.; Peterson, R.Larry.
> Distribution of foliar fungal endophytes of *Pinus strobus* between and within host trees. Canadian Journal of Forest Research; Sep 01 2000. Description: The distribution of foliar fungal endophytes within and between needles and trees of *Pinusstrobus* L. (white pine) is largely unknown. In this study, needles were collected in Muskoka, Ontario, plated, and scored for hyphal outgrowth of endophytes to observe distributional patterns. Individual trees displayed different levels of infection but branches within those trees had similar levels. Vol. 30; no. 9; pp. 1436-1442

Dehayes, D.
> Freezing injury to montane Picea rubens: the potential role of and mechanisms for natural and anthropogenic perturbations Bulletin of the Ecological Society of America; Jun 30 1994. Description: Winter injury occurs frequently to current-year needles of red spruce in the northern Appalachians and appears to be caused by subfreezing temperatures rather than foliar desiccation. Under ambient conditions, the maximum depth of cold tolerance achieved by red spruce needles in midwinter is barely sufficient to avoid freezing injury at common winter temperatures. Therefore, any perturbation that would decrease midwinter cold tolerance by just a few degrees substantially increases the probability of freezing injury. Vol. 752; pp. 51-52

DeLonay, A.J.; Little, E.E.; Woodward, F.; Brumbaugh, W.G.; Farag, A.M.; Rabeni, C.F..
> Sensitivity of early-life-stage golden trout to low pH and elevated aluminum Environmental Toxicology and Chemistry; Jul 31 1993. Description: Early-life-stage golden trout (Oncorhynchus aguabonita) were exposed to acid and Al to examine the response and determine the sensitivity of a western, alpine salmonid to conditions simulating an episodic pH depression. Freshly fertilized eggs, alevins, and swim-up larvae were exposed for 7 d to one of 12 combinations of pH and Al, and surviving fish were held to 40 d post-hatch to determine the effect of exposure on subsequent survival and recovery. Golden trout are sensitive to conditions simulating episodic acidification events typically

observed in the field. Significant mortality occurred when the pH of test waters was below 5.0 in the absence of Al or when pH was 5.5 in the presence of 100 [mu]g/L total Al. Behavioral impairments were sensitive indicators of low pH and Al stress. Vol. 127; pp. 1223-1232

Desai, M.S.; King, D.J.; Elgawhary, A.M. Clean Air Act Amendments and their impacts. Proceedings of the American Power Conference; Jan 31 1991. Description: Acid Deposition Control, Title IV of the Clean Air Act Amendments of 1990, requires reduction of SO_2 emissions from existing electric utilities with SO_2 emissions greater than 1.2 lb/MMBtu, reduction of NO_x emissions, and offsets for SO_2 emissions from new plants. The act permits emission trading for reducing SO_2. Emission trading permits utilities to select the most cost-effective method of compliance. This paper discusses the principal impacts of Acid Deposition Control and reviews compliance options. These options include coal switching and/or blending, addition of SO_2 controls, repowering, load management, conservation and use of renewable energy, and emission trading. Vol. 53; pp. 339-344

Devitt, T.W.; Weinstein, D.M. Acid rain mitigation: Everyone benefits from a market for compliance. Public Utilities Fortnightly; Mar 01 1990. Description: The writers of this article anticipate that there will be a market in credits for overcompliance with sulfur dioxide emission limitations created by new congressional legislation, and advocate that utility managers plan to participate actively in that market. In the article they use some actual data on the compliance options of utilities, and their costs, to determine a market-clearing price for compliance credits. They also proceed to show that every utility affected by the new law will be better off if it participates in the compliance market. Vol. 1255; pp. 14-18

Djuric, M; Ranogajec, J; Omorjan, R.; Miletic, S.. Sulfate corrosion of Portland cement -- Pure and blended with 30% of fly ash Cement and Concrete Research; Sep 30 1996. Description: This paper considers the sulfate corrosion of Portland cement-pure and blended with 30 wt% of fly ash. As raw materials, three kinds of Portland cements, with different content of C_3A (5.45--11.84 wt%), were applied. Fly ash was added so as to decrease the C_3A content (to 3.82--8.29 wt%). The test samples were exposed to the influence of aggressive environment (solutions with 1,000 and 2,000 mg/l of the So_4^{-2} ions as well as 320 mg/l of the NH_4^{-2} ions). Experiments lasted six months, while the flexural strengths were measured after 1, 2, 4 and 6 months. The corrosion resistance factor was calculated. For the examined systems, a correlation was suggested. It expresses the resistance factor as a function of three independent variables (the C_3A content, duration of corrosion process and concentration of the SO_4^{-2} ions). Vol. 26; no. 9; pp. 1295-1300

Dobson, J.E.; Rush, R.M.; Peplies, R.W. Forest blowdown and lake acidification Association of American Geographers, Annals; Jan 31 1990. Description: The authors examine the role of forest blowdown in lake acidification. The approach combines geographic information systems (GIS) and digital remote sensing with traditional field methods. The methods of analysis consist of direct observation, interpretation of satellite imagery and aerial photographs, and statistical comparison of two geographical distributions-one representing forest blow-down and another representing lake chemistry. Spatial and temporal associations between surface water pH and landscape disturbance are strong and consistent in the Adirondack Mountains of New York. In 43 Adirondack Mountain watersheds, lake pH is associated with the percentage of the watershed area blown down and with hydrogen ion deposition. Vol. 803; pp. 343-361

Dragovich, D.
> Marble weathering in an industrial environment, eastern Australia Environmental Geology and Water Sciences; Jan 31 1991. Description: Qualitative evidence from monuments and buildings in industrialized countries indicates that rates of stone deterioration rise in the presence of urban and industrial pollutants. Measurements presented here on surface reduction of marble tombstones show that mean weathering rates have increased over the period 1885 and 1955. Weathering rates were lower before the establishment of sulfur dioxide- emitting plants. Vol. 172; pp. 127-132

Drdla, E.
> Electrostatic precipitator electrode upgrade for low sulfur western coal at the OPPD Nebraska City station. Proceedings of the American Power Conference; Jan 31 1992 Description: The Omaha Public Power District (OPPD) Nebraska City Unit 1 coal fired boiler (585 MW) began operation in 1980. It was quickly recognized that the electrostatic precipitator would be a major detriment to the unit's availability. After nine years of boiler availability averaging approximately 70% due to electrostatic precipitator forced outages, a major upgrade was undertaken in 1989. Vol. 541; pp. 53-59

Dreschel, T.W.; Madsen, B.C.; Maull, L.A.; Hinkle, C.R.; Knott, W.M.
> Precipitation chemistry: Atmospheric loadings to the surface waters of the Indian River lagoon basin by rainfall. Florida Scientist; Jan 31 1990. Description: Rain volume and chemistry monitoring as part of the Kennedy Space Center Long Term Environmental Monitoring Program included the years 1984-1987 as part of the National Atmospheric Deposition Program. Atmospheric deposition in rainfall consisted primarily of seasalt and hydrogen ion, sulfate, nitrate, and ammonium ions. The deposition of nitrogen was on the order of 200-300 metric tons per year to the surface waters. Vol. 533; pp. 184-188

Drexhage, Michael.; Gruber, Franz.
> Architecture of the skeletal root system of 40-year-old (Germany). Canadian Journal of Forest Research; Jan 01 1998. Description: Altogether 15 root systems, five at each of three plots (north- and south-facing slopes and plateau), of 40-year-old *Picea abies* (L.) Karst. trees with different symptoms of forest decline were excavated down to a root diameter of 0.5 cm. The object was to investigate the variability of root morphology and to assess the influence of environmental variation on the architecture of the woody root system. For each tree, total height, diameter at breast height, and needle and twig biomasses were determined, and for each root system, biomass, growth, length, cross-sectional area, number and initial direction of branches, and branching forms were determined. Vol. 28; no. 1; pp. 13-22

Driscoll, C.T.
> Environmental chemistry of small watersheds. American Chemical Society, Division of Environmental Chemistry, Preprints; Jan 31 1990. Description: The small watershed approach is widely used to address a range of environmental chemistry questions. These may pertain to the effects of atmospheric deposition of pollutants (e.g., acid rain, trace metals), biogeochemical processes occurring within catchments (e.g., weathering, hydrologic flow paths) or land-use issues (e.g., clear-cutting practices, sustainability). An overview presentation will be given for the session on the Environmental Chemistry of Small Watersheds. The lecture will be directed towards issues for discussion during the session. Vol. 301; pp. 183

Dudek, D.J.
> Energy and environmental policy: The role of markets. Natural Resources and Environment; Jan 31 1991. Description: This article discusses the conflicts created by the fact that while the drive for economic development has been fueled by both resource exploitation and strategic imperatives, energy policy over the last twenty years has been made largely by environmental regulators. The main topics

discussed include: remedies for market failure; acid rain and the new market in emissions allowances; using markets to reduce greenhouse-gas emissions.
Vol. 6; no. 2; pp. 22-25, 59

Dudek, D.J.; Goffman, J.
Can preapproval jump-start the allowance market. Electricity Journal; Jun 30 1992 Description: With compliance deadlines approaching in three years, utility, environmental and financial planners and their regulators are in the process of grappling with the requirements imposed, and opportunities created, by the acid rain program established under Title 4 of the Clean Air Act amendments of 1990. The novel element of the program - emissions or allowance trading through a nationwide allowance market - presents great challenges for utilities and their regulators. Perhaps the foremost challenge is establishing the allowance market.
Vol. 55; pp. 12-17

Editorial: Acid precipitation
Environment International; Jan 31 1995 Description: This editorial focuses on acid rain and the history of public and governmental response to acid rain. Comments on a book by Gwineth Howell 'Acid Rain and Acid Waters' are included. The editor feels that Howells has provide a service to the environmental scientific community, with a textbook useful to a range of people, as well as a call for decision makers to learn from the acid rain issue and use it as a model for more sweeping global environmental issues. A balance is needed among several parameters such as level of evidence, probability that the evidence will lead to a specific direction and the cost to the global community. Vol. 21; no. 4; pp. 351-352

Edwards, R.A.; Blake, M.W.; Burris, L.H. Jr.; Carpenter, R.T.
Production capital project analysis for the Southern Company. Proceedings of the American Power Conference; Jan 31 1992 Description: Capital budgeting decisions, more than anything else, determine the future of an electric utility. The 1992 capital budget for The Southern Company exceeds $1 billion. By 1994, the annual capital budget is expected to double, and by 1998, it is expected to triple as a result of acid rain compliance and power plant additions. A sizable portion of the capital budget is for production projects to improve existing power plant performance. Analysis of the net present-value benefits of these projects is essential in making good capital budgeting decisions. This paper describes the process used by The Southern Company to analyze production capital projects that exceed $50,000.
Vol. 542; pp. 918-924

Ek, A.S.; Korsman, T.
A paleolimnological assessment of the effects of post-1970 reductions of sulfur deposition in Sweden. Canadian Journal of Fisheries and Aquatic Sciences; Aug 01 2001. Description: Analysis of diatoms in sediment cores from 10 acidic (pH < 6) lakes in southern Sweden shows that eight of the lakes have acidified after 1950, while two lakes have not significantly acidified. However, since the 1970s, sulfur deposition has decreased by 50%, and lake water chemistry monitored since 1983 shows an initial reversal of acidification. However, the diatom data do not indicate that a general recovery in pH has occurred yet. Vol. 58; no. 8; pp. 1692-1700

Elless, M.P.; Armstrong, A.Q; Lee, S.Y.
Characterization and solubility measurements of uranium-contaminated soils to support risk assessment. Health Physics; May 31 1997. Description: Remediation of uranium-contaminated soils is considered a high priority by the US Department of Energy because these soils, if left untreated, represent a hazard to the environment and human health. Because the risk to human health is a function of the solubility of uranium in the soils, the objectives of this work are to measure the uranium solubility of two contaminated soils, before and after remedial treatment, and determine the health risk associated with these soils. Two carbonate-rich, uranium-contaminated soils from the US Department of Energy Fernald

Environmental Management Project facility near Cincinnati, Ohio, as well as two nearby background soils were characterized and their uranium solubility measured in a 75-d solubility experiment using acid rain, groundwater, lung serum, and stomach acid simulants.
Vol. 72; no. 5; pp. 716-726

Ellis, H.; Bowman, M.L.
Critical loads and development of acid rain control options. Journal of Environmental Engineering (New York); Jan 31 1994 Description: The present paper describes the result of a demonstration project in which optimization models were used to identify efficient and equitable SO_2 control strategies that satisfy environmental quality standards at sensitive receptor locations in Maryland. Notable in the modeling is the use of location- specific critical-load estimates. These estimates -- the critical loads -- represent maximum allowable sulfur-deposition rates, that is, rates below which continued acidification and its attendant possible ecological consequences are believed not to occur. The analyses include scenarios that involve 10,000,000 t/yr SO_2 reductions in the United States, uniform percentage-reduction scenarios, and state-level sensitivity analyses.
Vol. 1202; pp. 273-290

Ember, L.R.
Clean air law will be costly to chemical industry. Chemical and Engineering News; Nov 12 1990. Description: The author discusses the impact of the 1990 revision to the Clean Air Law on the chemical industry. The revision seeks to stem smog, acid rain, and air pollutants at their sources. Vol. 6846; pp. 23-24

EPA reports results of first SO_2 allowance auction. Air Pollution Consultant; Jan 31 1993. Description: SO_2 allowances are the product of an innovative effort to control SO_2 emissions as part of the acid rain program (Clean Air Act Title IV and 40 CFR Part 73). Each allowance permits its holder to emit one ton of SO_2 during a specified year. Allowances can be bought, sold, and traded. An allowance reserve, which consists of 2.8% of the total allowance allocated to utilities nationwide, is withheld by EPA for the purpose of stimulating and supporting on the allowance market through both auctions and direct sales. In addition, other parties holding allowances can offer them for sale at annual auctions. Vol. 35; pp. 2.37-2.38

Evans, A. Jr.; Zelazny, L.W.
Kinetics of aluminum and sulfate release from forest soil by mono- and diprotic aliphatic acids. Soil Science; Jun 30 1990. Description: A batch equilibration study evaluated the influence of naturally occurring low- molecular-weight mono- and diprotic aliphatic acids on the rate of Al and SO_4^{2-} release in a Cecil soil (Typic Hapludult). The authors adjusted the pH of the organic acids (OAs) and of the soil suspension (3.8% w/w) to pH 4.0 and allowed them to equilibrate thermally before the experiment. After rapid addition of OAs to the soil suspension, they took solution samples at various time intervals and analyzed for Al, SO_4^{2-}, and OA concentration. The initial concentration of OA in suspension was 1×10^{-5} mol liter^{-1}. Both Al and SO_4^{2-} release followed pseudo- first-order kinetics, whereas OA adsorption obeyed simple first-order kinetics. The rate of Al release (k_1) was more rapid for the diprotic OA treatment (20.4×10^{-8} mol s^{-1}), as was SO_4^{2-} release (1.63×10^{-8} mol s^{-1}), compared to the monoprotic OA treatment. The rate of Al release varied inversely with OA chain length and the distance between -COOH functional groups. The addition of substituent -OH groups between the -COOH groups further reduced K_1. A similar trend was observed for the rate of SO_4^{2-} release (k_1) into solution. Monoprotic OAs were more rapidly adsorbed to the particle surfaces than were diprotic OAs. The authors postulate that removal of Al and SO_4^{2-} from solution occurs via selective mineral precipitation.
Vol. 1496; pp. 324-330

Fahrer, S.
Emissions trading programs, making sense of the options. Chemical Engineering (New

York); Mar 31 1996. Description: In an attempt to move away from the traditional command-and-control approach to regulation, the US Environmental Protection Agency has begun to develop economic incentive programs. These programs encourage compliance with nationwide pollution-reduction goals, but seek industry action based on market or profit incentives, rather than fear of retribution or penalty. The 1990 Clean Air Act Amendments (CAAA) require that stringent means be taken to reduce NO_x pollution in so-called ozone-nonattainment areas. Vol. 103; no. 3; pp. 139-144

Farag, A.M.; Woodward, D.F.; Little, E.E.; Steadman, B.; Vertucci, F.A.
The effects of low pH and elevated aluminum on yellowstone cutthroat trout (Oncorhynchus clarki bouvieri)
Environmental Toxicology and Chemistry; Apr 30 1993
Description: Although acid deposition is not considered a problem in the western US, surface waters in high elevations and fish inhabiting these waters may be vulnerable to acidification. This study examined the sensitivity of a wester salmonid to acid and aluminum stress. Yellowstone cutthroat trout (Oncorhynchus clarki bouvieri; YSC) were exposed for 7 d during each of four early life stages, or continuously from fertilization to 40 d post-hatch, to decreased pH and elevated Al.
Vol. 124; pp. 719-731

Faust, B.C.; Anastasio, C.; Allen, J.M.; Takemitsu, A.
Aqueous-phase photochemical formation of peroxides in authentic cloud and fog waters. Science (Washington, D.C.); Apr 02 1993. Description: Gas-to-drop partitioning of hydrogen peroxide and its precursor, the hydroperoxyl radical ($HO_2^\bullet$), has been considered the predominant or sole source of hydrogen peroxide in atmospheric water drops. However, atmospheric water can absorb solar ultraviolet radiation, which initiates the photoformation of peroxides (primarily hydrogen peroxide). Measurements of peroxide photoformation rates in authentic atmospheric water samples demonstrate that aqueous-phase photochemical reactions are a significant, and in some cases dominant, source of hydrogen peroxide to cloud and fog drops.
Vol. 2605104; pp. 73-75

Fenn, M.E; Leininger, T.D
Uptake and distribution of nitrogen from acidic fog within a ponderosa pine (Pinus ponderosa Laws.)/litter/soil system
Forest Science; Nov 30 1995
Description: The magnitude and importance of wet deposition of N in forests of the South Coast (Los Angeles) Air Basin have not been well characterized. We exposed 3-yr-old ponderosa pine (Pinus ponderosa Laws.) seedlings growing in native forest soil to acidic fog treatments (pH 3.1) simulating fog chemistry from a pine forest near Los Angeles, California. Fog solutions contained either $^{15}NH_4^+$, $^{15}NO_3^-$, or unlabeled N. The fog treatments were applied in open-top chambers in six 5-hr exposures. Soil treatments within each of the fog exposures were bare soil, soil overlain with L- and F-litter, and soil covered with plastic during the fog events to prevent fogwater from contacting soil.
Vol. 41; no. 4; pp. 645-663

Fey, M.V.; Schuette, R.; Manson, A.D.
Acidification of the pedosphere
South African Journal of Science; Jan 31 1990. Description: Acids accumulate naturally in soil through the processes of hydrolysis and leaching of base cations. Acidification is greatly intensified by anthropogenic distortions of the nitrogen and sulfur cycles, and by afforestation and the export of bases in harvested products. The rate of soil acidification is potentially highest in agriculture, where it's associated mainly with liberal use of ammoniacal fertilizers and production of forage legumes. Vol. 8678910; pp. 403-406

Fiss, F.C.; Carline, R.F.
Survival of brook trout embryos in three episodically acidified streams.
Transactions of the American Fisheries

Society; Mar 31 1993. Description: We evaluated, for brook trout Salvelinus fontinalis in three streams that undergo episodic acidification during critical periods of embryo development, survival of embryos from egg disposition to preemergence in natural redds and survival of sac fry in toxicity tests done in situ. Twenty-five natural redds were used for comparisons among streams.
Vol. 1222; pp. 268-278

Fitzgerald, W.F.; Clarkson, T.W.
Mercury and monomethylmercury: Present and future concerns. Environmental Health Perspectives; Dec 31 1991. Description: Global atmospheric changes carry the potential to disrupt the normal cycling of mercury and its compounds. Acid rain may increase methylmercury levels in freshwater fish. Global warming and increased ultraviolet radiation may affect the global budget of methylmercury, including its formation and degradation in both biotic and abiotic environments.
Vol. 96; pp. 159-166

Flanigan, T.
The 24 benefits of energy efficiency to electric utilities. Cogeneration and Competitive Power Journal; Jan 31 1995. Description: While electric utilities in the United States brace themselves for a more competitive era, many advocates of energy efficiency are concerned that such an era may exclude utilities' new-found emphasis on energy efficiency programs. Utilities are "right-sizing" (cutting staff) and reducing their operating budgets in every way possible to stabilize if not lower their rates. all as a means of remaining competitive and financially viable. Demand-side management programs appear at risk for they too have rate impacts. Several utilities with prominence in the delivery of customer energy efficiency services (DSM) strategies and that they will likely cut their DSM budget dramatically. Vol. 10; no. 2; pp. 59-65

Flaten, T.P.; Alfrey, A.C.; Birchall, J.D.
Status and future concerns of clinical and environmental aluminum toxicology Journal of Toxicology and Environmental Health; Aug 30 1996. Description: A wide range of toxic effects of aluminum (Al) have been demonstrated in plants and aquatic animals in nature, in experimental animals by several routes of exposure, and under different clinical conditions in humans. Aluminum toxicity is a major problem in agriculture, affecting perhaps as much as 40% of arable soils in the world. In fresh waters acidified by acid rain, Al toxicity has led to fish extinction. Aluminum is a very potent neurotoxicant. In humans with chronic renal failure on dialysis, Al causes encephalopathy, osteomalacia, and anemia. There are also reports of such effects in certain patient groups without renal failure.
Vol. 48; no. 6; pp. 527-541

Fleyfel, F.; Richardson, H.H.; Devlin, J.P.
Comparative SO_2 infrared spectra: Type I and II clathrate hydrate films, large gas-phase clusters, and anhydrous crystalline films. Journal of Physical Chemistry; Sep 06 1990. Description: The mechanism by which SO_2 is incorporated into microparticles of ice in the vapor phase is receiving special interest because of the unexpectedly high efficiency with which SO_2 is scavenged by ice crystals. A possible explanation of this efficiency might be found in the tendency for small polar molecules, such as the small ring ethers, to form clathrate hydrates at low temperatures and low partial pressures. This possibility has been examined by spectroscopic studies at ~ 120 K of large gas-phase clusters formed from anhydrous SO_2 and H_2O-SO_2 mixtures with a ratio appropriate for clathrate hydrate formation.
Vol. 9418; pp. 7032-7037

Forti, M.C.; Moreira-Nordemann, L.M.
Rainwater and throughfall chemistry in a terra firme rain forest: Central Amazonia Journal of Geophysical Research; Apr 20 1991. Description: During the Global Tropospheric Experiment (GTE)-Amazon Boundary Layer Experiment (ABLE) 2B campaign in the Amazon basin, samples of rainwater and throughfall were obtained in a terra firme (nonflooded forest) rain forest

at the Ducke Reserve (2° 57{prime}S, 59° 58{prime}W). The samples were collected during one wet period (April 1 to May 13, 1987) and one dry period (August 1 to October 1, 1987). All samples were analyzed for Na^+, K^+, Mg^{2+}, Ca^{2+}, NH_4^+, Cl^- and SO_4^{2-}, and pH. The rainwater was acidic, with a volume-weighted mean pH of 4.6 for two periods. Rainwater input from the dry period was 2 times greater for Na^+, Mg^{2+}, NH_4^+ and SO_4^{2-} and about 4 times greater for K^+ than from the wet period. The ionic concentrations in throughfall were higher than those in rainwater, except for NH_4^+ during the dry period. This enrichment of throughfall is attributed to the interaction of precipitation with the forest canopy.
Vol. 964; pp. 7415-7421

Fournier, R.E.; Morrison, I.K.; Hopkin, A.A.
Short range variability of soil chemistry in three acid soils in Ontario, Canada
Communications in Soil Science and Plant Analysis; Jan 31 1994. Description: The objective of this study was to assess the efficacy of soil sampling and analysis methodologies used in Canada's Acid Rain National Early Warning System (ARNEWS). During July and August of 1992, twenty-five soil pits were sampled and analyzed for available phosphorus (P); exchangeable potassium (K), calcium (Ca), magnesium (Mg), iron (Fe), copper (Cu), manganese (Mn), zinc (Zn), aluminum (Al), sodium (Na), sulfate-sulfur (SO4-S), boron (B), and molybdenum (Mo); total nitrogen (N), P, K, Ca, Mg, Fe, Cu, Mn, Zn, Al, Na, sulfur (S), B, nickel (Ni), lead (Pb), and organic matter (O.M.); pH; and cation exchange capacity (C.E.C.) at three ARNEWS sites across Ontario.
Vol. 251718; pp. 3069-3082

Freedman, B.; Zobens, V.; Hutchinson, T.C.; Gizyn, W.I.
Intense, natural pollution affects Arctic tundra vegetation at the Smoking Hills, Canada. Ecology; Apr 30 1990.
Description: Long-term, natural emissions of sulfur dioxide and acidic aerosols have had an impact on remote tundra at the Smoking Hills. The emissions have caused plant damage by SO_2 toxicity, and have severely acidified soil and freshwater. At the most intensively fumigated locations closest to the sources of emission, pollution stresses have devegetated the terrestrial ecosystem.
Vol. 712; pp. 492-503

Friedmann, A.S; Leiter, J.C.; Watzin, M.C.
Effects of environmental mercury on gonadal function in Lake Champlain northern pike (Esox lucius)
Bulletin of Environmental Contamination and Toxicology; Mar 31 1996
Description: Levels of mercury in the environment have increased steadily over the past two centuries, primarily because of human activity. Common point sources of this heavy metal include industrial waste discharge from chloralkali and paper pulp plants. More diffuse emissions, which become widely distributed by global wind currents, result from the combustion of fossil fuels and incineration of municipal wastes. Stricter laws in the United States have decreased the amount of pollution from point sources. In contrast, mercury from diffuse atmospheric origins has been increasing, causing a rise in rainwater concentrations and aquatic environments frequently distant from the source of pollution. Vol. 56; no. 3; pp. 486-492

Frisbie, M.P.; Wyman, R.L.
A field test of the effect of acidic rain on ion balance in a woodland salamander
Bulletin of the Ecological Society of America; Jun 30 1994. Description: Earlier laboratory studies demonstrated that red-backed salamanders, Plethodon cinereus, are susceptible to osmotic disruption by low pH substrates. In natural systems, however, acidic input from precipitation may be mediated by soils before it impacts salamanders. We tested the effect of acidic rain on sodium balance in salamanders by confining individuals in enclosure in two forest types (hemlock, beech) for 34 d. Enclosures received artificial rain of either pH 3 or 5 every 3- 4 d. Soils inside enclosures in the hemlock forest were more acidic than those in the beech forest at the outset. Vol. 752; pp. 71

Fu, Ji-Meng; Winchester, J.W.
Inference of nitrogen cycling in three watersheds of northern Florida, USA, by multivariate statistical analysis Geochimica et Cosmochimica Acta; Mar 31 1994. Description: Nitrogen in fresh waters of three rivers in northern Florida- the Apalachicola- Chattahoochee-Flint (ACF) River system, Ochlockonee (Och), and Sopchoppy (Sop)- is inferred to be derived mostly from atmospheric deposition. Because the N:P mole ratios in the rivers are nearly three times higher than the Redfield ratio for aquatic photosynthesis, N is saturate in the ecosystems, not a limiting nutrient, although it may be chemically transformed. Absolute principal component analysis (APCA), a receptor model, was applied to many years of monitoring data for Apalachicola River water and rainfall over its basin in order to better understand aquatic chemistry of nitrogen in the watershed. Vol. 586; pp. 1591-1600

Fu, Ji-Meng; Winchester, J.W.
Sources of nitrogen in three watersheds of northern Florida, USA: Mainly atmospheric deposition. Geochimica et Cosmochimica Acta; Mar 31 1994. Description: Atmospheric deposition is estimated to be the principal source of N in water that flows to the Apalachicola river from the Chattahoochee and Flint Rivers (ACF) as well as in two nearby small rivers, Ochlockonee (Och) and Sopchoppy (Sop), that drain watersheds with different land use characteristics. By mass balance and descriptive statistics of hundreds of rainfall and river water samples from monitoring programs since the 1960s, the average nitrate and ammonium deposition flux from the atmosphere is sufficient to account for N that flows toward Apalachicola Bay, an estuary in which N may be a limiting nutrient. Urban and agricultural sources of N in the three watersheds ACF, Och, and Sop appear to be relatively smaller.
Vol. 586; pp. 1581-1590

Gagen, C.J.; Sharpe, W.E.; Carline, R.F.
Mortality of brook trout, mottled sculpins, and slimy sculpins during acidic episodes Transactions of the American Fisheries Society; Jul 31 1993. Description: Brook trout Salvelinus fontinalis, mottled sculpins Cottus bairdi, and slimy sculpins Cottus cognatus occur in many Pennsylvania streams that have depressed pH and elevated aluminum concentrations during episodes of high stream discharge (acidic episodes). We performed 20-d in situ cage exposures with these species to determine their relative sensitivities to field conditions. We also exposed fish in the laboratory to synthetic soft water, without added Al, to elevate possible effects of Al on sodium flux rates and pH toxicity.
Vol. 1224; pp. 616-628

Galloway, J.N.
Anthropogenic mobilization of sulfur and nitrogen: Immediate and delayed consequences. Annual Review of Energy and the Environment; Jan 31 1996. Description: Global mobilization and dispersal of sulfur (S) and nitrogen (N) have been significantly increased by human activities. They are projected to increase even more in the future owing to growth in population and per-capita consumption of food and energy in the developing world, primarily Asia. Increased mobilization and distribution result in changes in precipitation acidity, ecosystem alkalinity and nutrient status, tropographic and stratospheric ozone concentrations, and energy balance of the troposphere. Vol. 21; pp. 261-292

Garten, C.T. Jr.
Foliar leaching, translocation, and biogenic emission of ^{35}S in radiolabeled loblolly pines. Ecology; Feb 28 1990. Description: Foliar leaching, basipetal (downward) translocation, and biogenic emission of sulfur (S), as traced by ^{35}S, were examined in a field study of loblolly pines. Four trees were radiolabeled by injection with amounts of ^{35}S in the 6-8 MBq range, and concentrations in needle fall, stemflow, throughfall, and aboveground biomass were measured over a period of 15-20 wk after injection. The contribution of dry deposition to sulfate-sulfur (SO_4^{2-}-S)

concentrations in net throughfall (throughfall SO_4^{2-}-S concentration minus that in incident precipitation) beneath all four trees was > 90%. Calculations indicated that about half of the summertime SO_2 dry deposition flux to the loblolly pines was fixed in the canopy and not subsequently leached by rainfall. Based on mass balance calculations, ^{35}S losses through biogenic emissions from girdled trees were inferred to be 25-28% of the amount injected. Estimates based on chamber methods and mass balance calculations indicated a range in daily biogenic S emission of 0.1-10 {mu}g/g dry needles. Translocation of ^{35}S to roots in nongirdled trees was estimated to be between 14 and 25% of the injection. It is hypothesized that biogenic emission and basipetal translocation of S (and not foliar leaching) are important mechanisms by which forest trees physiologically adapt to excess S in the environment. Vol. 711; pp. 239-251

Gauri, K.L.; Punuru, A.R.; Holdren, G.C.
Acidity gradients in the KIPDA region Environmental Geology and Water Sciences; Jan 31 1990. Description: Acid precipitation leaves fingerprints upon marble surfaces. Preserved in weathering crusts on dated monuments, these fingerprints serve as guides to identify levels of acidity of a region. The authors collected scrapings from protected surfaces of tombstones made of Georgia marble whose dates of installation could be determined with reasonable certainty. Vol. 151; pp. 55-58

Geissler, M.; Eldik, R. van
Development of a gradient ion-pair chromatographic procedure for the simultaneous detection of nitrogen-sulfur oxides produced during the reaction of SO_x and NO_y species in aqueous solution Analytical Chemistry (Washington); Dec 01 1992. Description: The autoxidation reactions of sulfur and nitrogen oxides play an important role in atmospheric chemistry in terms of acid rain formation. An ion-pair chromatographic procedure using an acetonitrile gradient was developed for the simultaneous detection of a series of nitrogen-sulfur oxides produced during the reaction of nitrogen and sulfur oxides. These oxides include nitrilotrisulfonate (NTS), imidodisulfonate (IDS), hydroxylaminedisulfonate (HADS), N-nitrosohydroxylamine-N-sulfonate (NHAS), hydroxylaminesulfonate (HAMS), and aminosulfonate (SA). Detection limits that could be reached were in the ppb range. Vol. 6423; pp. 3004-3006

Ghosh, Mini; Chandra, Peeyush; Sinha, Prawal
A Mathematical Model to Study the Effect of Toxic Chemicals on aPrey-Predator Type Fishery. J. Bio. Syst; Jun 01 2002 Description: In this paper, a nonlinear mathematical model is proposed and analyzedto see the indirect effect of air pollutants on the prey-predator typefish population in a closed population (lake). It is shown that asthe pollutant concentration in the environment increases, theconcentration of the acidic chemicals in the lake increases andconsequently the equilibrium level of the fish population decreases.Using stability theory of differential equations and computersimulation, it is shown that due to the effort, pollutantconcentration can be reduced to a desired level to save fisheries fromextinction by acid rain. Vol. 10; no. 02

Ghuman, G.S.; Raut, K.B.
Vertical leaching of metals from sandy soil minerals in aqueous systems Georgia Journal of Science; Jan 31 1990 Description: Leaching studies were conducted on sandy soils from four plots in Coastal Georgia. Two plots were under grass and two were under forest vegetation. The objective of this study was to determine vertical release and transport of major and trace metals. Six successive leachings of soil columns packed to 36 cm depth, were carried out with two aqueous solutions, (1) distilled-deionized (DD) water acidified to pH 6.5, and (2) carbonated DD water (pH 5.0). All leachates were analyzed for the concentrations of metals using atomic absorption spectroscopy. Results indicated that both leaching solutions were equally

effective in the release of major metals, which were in a decreasing order as: Ca, Mg, K and Na in the grass-covered soil, and as: K, Na, Ca and Mg in the forested soil. The amounts of Ca, Mg and Na decreased in the sequential leachates, but those of K remained constant, indicating that coastal sandy soils are rich in K-feldspars. In the soils of all four plots, the concentrations of leached trace metals were in a decreasing order as: Ni, Zn, Mn, Cd and Cu with DD water (pH 6.5), and as: Mn, Zn, Ni, Cu and Cd with carbonated DD water (Ph 5.0). Results of this study simulate the effects of acid rain on the release and leaching of metals from surface soil downward into the groundwater. Vol. 481; pp. 39-40

Giamello, E.; Murphy, D.; Magnacca, G.; Morterra, C.; Shioya, Y.; Nomura, T.; Anpo, M. The interaction of NO with copper ions in ZSM5: An EPR and IR investigation. Journal of Catalysis; Aug 31 1992. Description: The interaction of nitric oxide with copper ZSM5 zeolites at room temperature has been studied be EPR and FT-IR spectroscopy in the aim of investigating the surface intermediates involved in the decomposition of NO to N_2 and O_2. Particular care has been devoted to obtaining a catalyst in a well-defined oxidation state, i.e., with one of the two ionic forms of copper (Cu^{2+} or Cu^+) clearly prevailing on the other. The interaction of NO with Cu^{2+}/ZSM5 yields a reversibly adsorbed nitrosylic adduct easily desorbed upon pumping at 333 K. The species is diamagnetic and bears a partial positive charge. Vol. 1362; pp. 510-520

Giesler, R; Moldan, F; Lundstroem, U. Reversing acidification in a forested catchment in southwestern Sweden: Effects on soil solution chemistry. Journal of Environmental Quality; Jan 31 1996. Description: The exclusion of acid precipitation in whole-catchment experiments can provide valuable information to further our understanding of recovery processes of acidified soils. In this study, we focused on the reversibility of acidification-induced changes in different soil horizons when anthropogenic deposition was excluded. A small forested catchment in the Gardsjoen area in southwest Sweden was covered with a transparent roof in April 1991 and sprinkled with water that simulated preindustrial deposition levels. Vol. 25; no. 1; pp. 110-119

Gillette, D.A.; Stensland, G.J.; Williams, A.L.; Barnard, W.; Gatz, D.; Sinclair, P.C.; Johnson, T.C. Emissions of alkaline elements calcium, magnesium, potassium, and sodium from open sources in the contiguous United States. Global Biogeochemical Cycles; Dec 31 1992. Description: Models of dust emissions by wind erosion (including winds associated with regional activity as well as dust devils) and vehicular disturbances of unpaved roads were developed, calibrated, and used to estimate alkaline dust emissions from elemental soil and road composition data. Emissions from tillage of soils were estimated form the work of previous researchers. The area of maximum dust production by all of those sources is the area of the old Dust Bowl' of the 1930s (the panhandles of Texas and Oklahoma, eastern New Mexico and Colorado, and western Kansas). Vol. 64; pp. 437-457

Gilliam, Frank.S.; Yurish, Bradley.M.; Adams, Mary.Beth. Temporal and spatial variation of nitrogen transformations in nitrogen-saturated soils of a central Appalachian hardwood forest Canadian Journal of Forest Research; Oct 01 2001. Description: We studied temporal and spatial patterns of soil nitrogen (N) dynamics from 1993 to 1995 in three watersheds of Fernow Experimental Forest, W.V.: WS7 (24-year-old, untreated); WS4 (mature, untreated); and WS3 (24-year-old, treated with $(NH_4)_2SO_4$ since 1989 at the rate of 35 kg $N \cdot ha^{-1} \cdot year^{-1}$). Net nitrification was 141, 114, and 115 kg $N \cdot ha^{-1} \cdot year^{-1}$, for WS3, WS4, and WS7, respectively, essentially 100% of net N mineralization for all watersheds. Vol. 31; no. 10; pp. 1768-1785

Gjessing, E.T.
 The HUMEX Project: Experimental acidification of a catchment and its humic lake. Environment International; Jan 31 1992. Description: Acid rain research during the late 1970s and the early 1980s concluded that acid precipitation seriously affected the environment. It was, however, realized that humic substances (HS) in the water have an effect on the response of acid rain, and that HS acts as a modifier on both the chemical composition and on the biological activity. The HUMEX Project is studing the impact of HS on the acidification and the effect acidification has on the biological properties of HS. Vol. 186; pp. 535-543

Global outreach
 Oak Ridge National Laboratory Review; Jan 31 1992. Description: The laboratories role in the 1990's can be characterized as much more global in scope. Many of the research efforts it has been involved in are now viewed as of concern on a global scale, environmental sciences being one example. The well developed expertise of the laboratory allows it to play a major role in the implementation and direction of research in these fields.
 Vol. 2534; pp. 235-267

Goldstein, B.D.; Reed, D.J.
 Global atmospheric change and research needs in environmental health sciences Environmental Health Perspectives; Dec 31 1991. Description: On November 6-7, 1989, the National Institute of Environmental Health Sciences (NIEHS) held a conference on Global Atmospheric Change and Human Health. As a result, and in the months since this conference, many important areas of research have been identified with regard to the impacts of climatic changes on human health. To develop comprehensive research programs that address important human health issues related to global warming, it is necessary to begin by recognizing that some of the health effects will be direct such as those due to temperature changes, and others will be indirect consequences of environmental alterations. Vol. 96; pp. 193-196

Gordon, D.
 Alternate fuels versus gasoline: A market niche. Forum for Applied Research and Public Policy; Jan 31 1994. Description: America travel on oil. Although many other fuels have the capability of moving the nation's cars, trucks, and buses, none have been able to carve out a significant market niche in the United States, notes Deborah Gordon of the Union of Concerned Scientists in Berkeley, California. The inherent advantages of gasoline as a motor fuel are well known: cost, availability, familiarity, ease of use, safety, and attractive physical properties, Gordon points out. Yet, its drawbacks also are significant: air pollution, global warming, acid rain, oil spills, groundwater contamination, and threats to national security and the nation's economic well being. Vol. 91; pp. 5-12

Graveland, Jaap.
 Effects of acid rain on bird populations. Environmental Reviews; Jan 01 1998. Description: In this paper the effects of anthropogenic acidification of soils and waters on bird populations are reviewed. Acidification causes (i) declines in the reproductive success and the density of piscivorous birds through declines in the fish populations, (ii) shifts in the forest bird community from forest birds to birds of open woodland through large-scale forest dieback, and (iii) leads to a lower reproductive success of birds in calcium-poor areas through a decline in the availability of calcium-rich material (needed for eggshell formation and skeletal growth). Vol. 6; no. 1; pp. 41-54

Grebenyuk, V.D.; Sobolevskaya, T.T.; Grebenyuk, O.V.; Konovalova, I.D.; Vysotskii, S.P.
 Electrochemical membrane conversion of sodium hydrogen sulfite for the purification of flue gases from sulfur dioxide. Journal of Applied Chemistry of the USSR (English Translation); Oct 20 1992. Description: One of the major ecological problems is the problem of acid rain. It can be solved by means of the new technology, an important element of which

is the conversion of sodium hydrogen sulfite into sodium sulfite and sulfur dioxide by electrodialysis with bipolar membranes. On the basis of experimental data on the electric conductivity of solutions and membranes and on the current efficiency under various conditions an equation is proposed for calculating the energy expenditures on the conversion of sodium hydrogen sulfite into sodium sulfite and sulfur dioxide.
Vol. 655; pp. 868-874

Greenberger, L.S.
Shopping for acid rain control strategies
Fortnightly; Jan 15 1992
Description: A utility manager trying to pick a compliance strategy today probably feels a lot like a child in a toy store: So many from which to choose - and so little time. This article examines the technologies available and which technologies utilities are likely to use to control SO_2 emissions.
Vol. 1292; pp. 37-40

Griffith, M.B; Perry, S.A; Perry, W.B.
Macroinvertebrate communities in headwater streams affected by acidic precipitation in the central Appalachians
Journal of Environmental Quality; Mar 31 1995. Description: We collected quantitative macroinvertebrate samples monthly from September 1989 to October 1990 from four streams on the Allegheny Plateau of West Virginia that were characterized by different bedrock geology and streamwater pH. Mean pH was 4.3, 6.1, and 6.0, and 7.5 in the four streams. We compared species and functional group composition of the benthic macroinvertebrate community in these streams to choose taxa that could be used as indicator species for differences in pH in bioassessment studies. The streams differed in species composition and abundance and several species were found that could be used as indicators for each of the levels of pH.
Vol. 24; no. 2; pp. 233-238

Grodzin´ska-Jurczak, M.; Godzik, B.
Air pollution and atmospheric precipitation chemistry in Poland - a review
Environmental Reviews; Oct 01 1999
Description: Poland, especially its southern and south-eastern parts, which are the most industrialized, has been exposed to enormous local and transboundary gaseous (SO_2, NO_x) and particulate (alkaline dusts, heavy metals) emissions. The fall of the communist system at the beginning of the 1980s and the economic changes that followed resulted in changes in industrial production, in the emission of pollutants, and in the pollution levels in the atmospheric precipitation. This paper describes these changes throughout Poland during the past 10 to 15 years.
Vol. 7; no. 2; pp. 69-79

Groffman, P. M.
Carbon additions increase nitrogen availability in northern hardwood forest soils. Biology and Fertility of Soils; Aug 03 1999. Description: Abstract The effects of acetate additions to northern hardwood forest soils on microbial biomass carbon (C) and nitrogen (N) content, soil inorganic N levels, respirable C and potential net N mineralization and nitrification were evaluated. The experiment was relevant to a potential watershed-scale calcium (Ca) addition that aims to replace Ca depleted by long-term exposure to acid rain.

Grupenhoff, J.T.
The case for a National Association of Physicians for the Environment
Environmental Research; Oct 31 1990
Description: Beleaguered by numerous public policy difficulties, public officials at all levels appear confused and bewildered when additionally faced with burgeoning environmental problems. Emotion and rhetoric often replace solid fact in much discussion about environmental policy development. And yet the environmental issues, global in scope in many cases (the ozone layer, global warming, acid rain), and those which are more locally differentiated as to impact (air pollution, water pollution, pesticides, occupational environmental threats), certainly could lend themselves to systematic scientific and medical scrutiny. When viewed this

way, it becomes clear that physicians' specialty organizations have an opportunity to make a major contribution to the development of environment public policy.
Vol. 531; pp. 1-5

Guinee, J.B; Heijungs, R.
A proposal for the definition of resource equivalency factors for use in product life-cycle assessment. Environmental Toxicology and Chemistry; May 31 1995 Description: Environmental life-cycle assessment (LCA) of products has been the focus of growing attention in the last few years. The methodological framework has been developed rapidly, and a provisional Code of Practice has been drawn up by an international group of experts. One of the elements of LCA is impact assessment, which includes a characterization step in which the contributions of resource extraction and polluting emissions to impact categories such as resource depletion, global warming, and acidification are quantified and aggregated as far as possible.
Vol. 14; no. 5; pp. 917-925

Günthardt-Goerg, Madeleine.S.; McQuattie, Carolyn.J.; Scheidegger, Christoph.; Rhiner, Claudia.; Matyssek, Rainer..
Ozone-induced cytochemical and ultrastructural changes in leaf mesophyll cell walls.
Canadian Journal of Forest Research; Apr 01 1997. Description: Cuttings of birch (Betula pendula Roth), poplar (Populus times euramericana (Dode) Guinier cv. Dorskamp), and alder (Alnus glutinosa (L.) Gaertn.) were exposed in the open field to ambient ozone (O_3), in both full sunlight and shade conditions, and in field fumigation chambers to filtered air (FA) or FA plus added O_3 (75 nL $\cdot L^{-1}$) from 07:00 to 19:00, 19:00 to 07:00, or for 24 h. Appearance of O_3-induced leaf symptoms was related to changes at the cellular level, especially in the cell wall.
Vol. 27; no. 4; pp. 453-463

Gunther, A.J.
A chemical survey of remote lakes of the Alagnak and Naknek river systems, southwest Alaska, U.S.A.. Arctic and Alpine Research (Boulder, Colorado); Feb 28 1992. Description: An analysis of the chemistry of 12 remote lakes of the Naknek and Alagnak drainages demonstrated significant differences in surface water alkalinity, due in part to high concentrations of naturally-occurring sulfate in the Alagnak drainage. The average surface water alkalinity for lakes in the Alagnak drainage is much less than that for the Naknek drainage, and below that for lakes from South Central Alaska, indicating a potential sensitivity of the lakes in the Alagnak river system to acid deposition. The upper Alagnak drainage contains Iron Springs Lake, a naturally acidic lake with a surface area of 70 ha. The acidity of this lake may be due to oxidation of surficial iron sulfide deposits, which would also explain the relatively high sulfate concentrations in the Alagnak drainage. Vol. 241; pp. 64-68

Gupta, G.; Krishnamurthy, S.
Changes in poultry litter toxicity with simulated acid rain. Bulletin of Environmental Contamination and Toxicology; Jan 31 1991. Description: The Delmarva Peninsula on the Eastern Shore of Maryland ranks 4th in the nation in poultry production and generates 9,500 metric tons of poultry manure/litter per day. The poultry litter contains many macro and micro nutrients and is an excellent source of fertilizer. The litter also contains antibiotics, heavy metals, hormones and many microorganisms. Land application of this litter has been the only means of its utilization and disposal. With rainfall, surface water run-off (leachate), from land on which litter has been applied, reaches the Cheasapeake Bay from this region. This leachate with its high organic and inorganic salt contents and high biochemical oxygen demand can severely disrupt the aquatic life and cause fish kills. The objective of this research was to study the effect of simulated acid rain (pH 3, 4 and 5) on the toxicity of poultry litter extracts. Vol. 461; pp. 167-172

Gurbin, G.M; Talbot, K.H
Nuclear hydrogen - cogeneration and the transitional pathway to sustainable development. Transactions of the American Nuclear Society; Jan 31 1994. Description: The atmospheric consequences of carbon of carbon and the evolution of world energy sources have resulted in a movement away from high carbon fuels, and a growing appreciation that the next generation of industrial development must be on a sustainable basis. The Bruce Energy Centre has been evolving for nearly two decades, driven by a mission to commercially demonstrate the importance of integrating energy, the environment and the economy in industrial development. The nearby Bruce Nuclear Generating Station "A" has provided process steam for operation of a fermentation alcohol plant, alfafa processing plant and fullscale greenhouse. Vol. 70; no. Suppl.1; pp. 169-176

Gureev, A.A.; Mitusova, T.N.; Sokolov, V.V.; Veretennikova, T.N.; Spirkina, N.P.; Melenchuk, A.I.
Improvement of ecological properties of diesel fuels. Chemistry and Technology of Fuels and Oils (English Translation); Jan 31 1993. Description: In the operation of high-speed diesel engines, the fuel composition and properties influence the toxicity and smoke level of the exhaust. Combustion of the sulfur compounds present in diesel fuel produces sulfur oxides that are responsible for metal corrosion, deterioration of roads and buildings, acid rain, and other undesirable phenomena. Particularly high concentrations of sulfur oxides are created in large cities where there is heavy automotive traffic. One of the principal means for reducing the content of sulfur oxides in the exhaust gas is a reduction of the sulfur content in the diesel fuel. 2 tabs. Vol. 2856; pp. 301-304

Guruswamy, L.D; Palmer, G.W.R. Si; Weston, B.H
International environmental law and world order. Energy Law Journal; Jan 31 1995. Description: A litany of dismal happenings - global warming, ozone layer depletion, desertification, destruction of biodiversity, acid rain, and nuclear and water accidents - are but some of the subjects covered by this book, a problem-solving casebook authored by three educators. This new book makes the obvious but important point, that environmental issues are not limited by national boundaries. The book is divided into three parts.
Vol. 16; no. 1; pp. 197-198

Hahn, R.W.; Stavins, R.N.
Economic incentives for environmental protection: Integrating theory and practice American Economic Review; May 31 1992 Description: For decades, economists have been extolling the virtues of market-based or economic-incentive approaches to environmental protection. Some 70 years ago, Arthur Cecil Pigou (1920) suggested corrective taxes to discourage activities that generate externalities. A half century later, J. Dales (1968) showed how the introduction of transferable property rights could work to promote environmental protection at lower aggregate cost than conventional standards
Vol. 822; pp. 464-468

Håkanson, L.; Andersson, T.; Nilsson, A.
New method of quantitatively describing drainage areas. Environmental Geology and Water Sciences; Jan 31 1990. Description: The aim was to introduce a new method, the DAZ method (drainage area zonation), to quantify environmental parameters, such as bedrocks, soil type, and land use in drainage areas. The work was carried out within the framework of the Swedish project Liming--mercury. Two important points in the project are that there are quantifiable relationships between the character of the drainage area and the lake and that several limnological and morphometric parameters may have an impact on the Hg content in fish. The DAZ method accounts for the fact that, for example, a certain soil type does not have an even distribution in the whole drainage area. Vol. 151; pp. 61-69

Håkanson, Lars
Assessment of critical loading of lakes as a basis for remedial measures: A review of fundamental concepts. Lakes & Reservoirs: Research and Management; Jan 31. Description: Abstract Lake ecosystems throughout the world are threatened by numerous chemicals. Most people have heard about the major chemical threats to aquatic systems, such as mercury, radionuclides, sulphur and acid rain, and nutrients causing eutrophication effects. How are these threats manifested in ecosystems? What is threatened and why? What can be done for remediation?
Vol. 6; no. 1; pp. 1-20

Halkos, G.E.
Economic incentives for optimal sulfur abatement in Europe. Energy Sources; Sep 30 1995. Description: This article reviews and develops theoretical and empirical representations of economic incentives for implementing pollution control strategies. A number of alternative economic instruments exist, which, if applied internationally, could encourage implementation of abatement strategies by counties. The article considers means of persuading countries to minimize abatement costs. A comparison between the pollution targets achieved by the imposition of a uniform charge rate and by differentiated charge rates is discussed, and empirical results are provided with associated conclusions. These results are compared with a simple standards setting in the form of critical loads, in order to assess empirically if economic instruments work better than regulations.
Vol. 17; no. 5; pp. 517-534

Halkos, G.E.
Evaluation of the direct cost of sulfur abatement under the main desulfurization technologies. Energy Sources; Jul 31 1995. Description: This study summarizes the available information on the technical characteristics and costs of those sulfur abatement technologies in operation at present or coming into operation in the near future. Relying on disaggregated source data and using engineering cost functions and various technical and economic assumptions, the least cost curves of sulfur abatement for all the European countries have been derived, and some examples are presented. Finally, a sensitivity analysis of abatement strategies and costs to some alternative assumptions about energy futures is presented.
Vol. 17; no. 4; pp. 391-412

Hallett, R.A.; Hornbeck, J.W.
Foliar and soil nutrient relationships in red oak and white pine forests. Canadian Journal of Forest Research; Aug 01 1997. Description: Red oak (Quercus rubra L.) and white pine (Pinus strobus L.) forests on sandy soils of the northeastern United States may have been depleted of nutrient cations by acid precipitation and intensive land use. Foliar Ca in oak was 5260 mg Ca roman kg^{-1} dry matter, or more than 3 times the amount in white pine foliage. Red oak also has more Mg, K, and N than white pine. Vol. 27; no. 8; pp. 1233-1244

Handley, C.O. Jr.
Terrestrial mammals of Virginia: Trends in distribution and diversity. Virginia Journal of Science; Jan 31 1991. Description: The present mammal fauna of Virginia formed during the post-Pleistocene warming trend. Indians had little impact on the fauna, but European introduction of firearms led to terminal exploitation of bison and elk and to deliberate extirpation of large predators. Logging, clearing for agriculture, and urbanization had a negative impact on some forest species and brought gains for some open country species. The present era of conservation attempts to maintain diversity and to stabilize the fauna through protection, restoration, and management.
Vol. 422; pp. 171

Hansen, B.K; Postma, D.
Acidification, buffering, and salt effects in the unsaturated zone of a sandy aquifer, Klosterhede, Denmark. Water Resources Research; Nov 30 1995. Description: Acidification of groundwater in a noncalcareous sandy aquifer at Klosterhede, Denmark, is the result of acid rain deposition. In the 4- to 5-m-thick

unsaturated zone the pH ranges from 4.2 to 4.9 with Al concentrations of up to 0.8 mmol L^{-1}. The groundwater at the top of the saturated zone still has a pH below 5. Deposition of sea salt affects the solute profiles, and its importance varies both spatially from the forest margin to the inner part of the forest and temporally through seasonal variations in infiltration and dry deposition.
Vol. 31; no. 11; pp. 2795-2810

Hansen, D.A.
CAMRAQ: Comprehensive regional air quality modeling. EPRI Journal (Electric Power Research Institute); Jan 31 1992. Description: As part of its long-standing research on regional air quality, Electric Power Research Institute (EPRI) has embarked on a project to develop a comprehensive modeling system for providing unique scientific guidance on emissions management options to industry and regulators. As the system evolves, interim versions can be used to address such problems as ozone nonattainment, visibility impairment, and acid rain. To leverage its investment and that of others in pursuing this goal, EPRI has fostered the formation of an international consortium of some 20 organizations - the Consortium for Advanced Modeling of Regional Air Quality, or CAMRAQ. Vol. 177; pp. 34-38

Hanson, B.M.
What happened to science
AAPG Bulletin (American Association of Petroleum Geologists); Sep 30 1993 Description: There is not a facet of anyone's life in industrial nations that is not related to us by hydrocarbons one way or another. Let us not decimate those industries that provide the infrastructure of our society by needless standards and regulations. Basic mineralogy would have told us that 95% of asbestos fibers are harmless,and would have save taxpayers millions of dollars in the so-called cleanup Vol. 779; pp. 1573

Hao, J.; Duan, L.; Zhou, X.; Fu, L.
Application of a LRT model to acid rain control in China. Environmental Science and Technology; Sep 01 2001. Description: For further control of acid rain and SO$_2$ pollution in China, acid rain control zones and sulfur dioxide pollution control zones were designated where acid rain or serious SO$_2$ pollution occurs or may occur. In this study, sulfur deposition in east China was computed through a policy-oriented, two-dimensional Eulerian model for long-range transport and deposition of SO$_2$ and SO$_4^{2-}$.
Vol. 35; no. 17; pp. 3407-3415

Hao, J.M.; Wang, S.X.; Liu, B.J.; He, K.B.
Designation of acid rain and SO$_2$ control zones and control policies in China. Journal of Environmental Science and Health, Part A: Toxic Hazardous Substances and Environmental Engineering; Jul 01 2000. Description: Effective SO$_2$ emission and acid rain controls are urgently needed in China. This paper designated the priority control zones for both SO$_2$ emission and acid precipitation in China. The Control Zones were identified as an area of 1.09×10^6 km^2, about 11.4% of the China's territory and about 14 million tons SO$_2$ emissions. The Acid Rain Control Zones had 806000 km^2, about 8.4% of the national territory and the SO$_2$ Pollution Control Zones had 290000 km^2, about 3% of the national territory. Vol. 35; no. 10; pp. 1901-1914

Harding, A.W; Brown, S.D; Thomas, K.M.
NO release from the isothermal combustion of coal chars. Preprints of Papers, American Chemical Society, Division of Fuel Chemistry; Jan 31 1996 Description: Coal combustion for power generation has associated environmental problems, in particular, the release of oxides of sulphur and nitrogen which are involved in the formation of acid rain. The modification and optimization of the combustion process to minimize the NO$_x$ emissions is therefore of considerable interest and importance. The combustion of coal occurs over two stages; (1) the rapid devolatilisation followed by combustion/ignition of the volatiles and (2) slower gasification of the residual char. The nitrogen present in the coal is partitioned between the volatile matter and

the residual char. Char nitrogen has been identified as the main contributor to NO emissions from low NO_x burners. Vol. 41; no. 1; pp. 165-169

Harding, A.W; Brown, S.D; Thomas, K.M.
Release of NO from the combustion of coal chars. Combustion and Flame; Dec 31 1996. Description: The emissions of nitrogen oxides during coal combustion contribute to acid rain and are a major environmental problem. In this investigation, the release of nitric oxide during the combustion of coal chars prepared from a wide range of coals in an entrained flow reactor was investigated over a range of combustion temperatures (823--1,323 K) using a thermogravimetric analyzer coupled with a mass spectrometer. The conversion of char nitrogen to NO (NO/char-N) was studied in relation to coal and char structural characterization parameters. The results show that higher levels of conversion of char-N to NO were observed for the high-rank coal chars at lower combustion temperatures, where the reaction is under chemical control. Vol. 107; no. 4; pp. 336-350

Hargeby, A.; Petersen, R.C. Jr.; Kullberg, A.; Svensson, M.
Benthic macroinvertebrates along the soil/water interface of the HUMEX lake 1989- 1991. Environment International; Jan 31 1992. Description: The taxonomic composition, abundance, and size distribution of benthic macroinvertebrates were studied at the soil/water interface two years before and the first year after the start of artificial acidification of a small catchment and its humic lake. The macroinvertebrate assemblage consisted mainly of predators; dragonflies (Odonata), damselflies (Zygoptera), net-building caddisflies (Polycentropodidae), diving beetles (Dytiscidae), and water bugs (Hemiptera). It is suggested that benthic and planktonic microcrustaceans are important prey for damselflies and that intraguild predation is important for the structure of the community. Vol. 186; pp. 659-666

Hargeby, A.; Stalhandske, P.; Svensson, M.; Kullberg, A.; Petersen, R.C. Jr.
Abundance, size distribution, and predation efficiency of Damselfly larvae (Zygoptera) in the HUMEX Lake Skjervatjern 1989-1992. Environment International; Jan 31 1994. Description: Damselfly (Zygoptera) larvae were studied in the littoral zone of the humic Lake Skjervatjern in western Norway. In 1988, the lake was divided by a plastic curtain into a control half (B) and an experimental half (A). Half A and its catchment have been treated with artificial acid rain since October 1990. Damselfly larvae were sampled in July of two consecutive years before (1989, 1990) and two after (1991, 1992) the start of the acidification. Vol. 203; pp. 343-348

Hart, G.S.
Sustainable air quality in the global commons. Clean Air (Melbourne); May 31 1990. Description: In this address, the speaker discusses ways in which a global commons approach to the development of national and international air pollution control policies may be advanced in relevant programs. He emphasizes the following considerations: public support for protection of the environment and a willingness to be taxed for the necessary financial support; the challenge to furtherance of globally sustainable development created by the changing political scene in the Soviet Union and Eastern Europe; the need for involvement of 3rd world countries in sustainable development; and threats to mankind from global warming, ozone depletion and acidification. Vol. 242; pp. 56-60

Hathaway, A.M. II
The environmental impact of energy efficiency. Strategic Planning for Energy and the Environment; Jan 31 1994. Description: Fast-paced federal stimulation has set new goals for energy and environmental improvements. The interrelationship between these two efforts grows more complex. An advanced and sophisticated process is needed: energy system efficiency integration. Vol. 133; pp. 46-59

Health effects of atmospheric acids and their precursors. Report of the ATS workshop on the Health Effects of Atmospheric Acids and their Precursors. American Review of Respiratory Disease (New York); Aug 31 1991. Description: Short communication. Vol. 1442; pp. 464-467

Hedin, L.O; Likens, G.E
Atmospheric dust and acid rain
Scientific American; Dec 31 1996
Description: Why is acid rain still an environmental problem in Europe and North America despite antipollution reforms? The answer really is blowing in the wind: atmospheric dust. These airborne particles can help neutralize the acids falling on forests, but dust levels are unusually low these days. In the air dust particles can neutralize acid rain. What can we do about acid rain and atmospheric dust? Suggestions range from the improbable to the feasible. One suggestion is to reduce emissions of acidic pollutants to levels that can be buffered by natural quantities of basic compounds in the atmosphere; this would mean continued reductions in sulfur dioxide and nitrogen oxides, perhaps even greater than those prescribed in the 1990 Amendments to the Clean Air Act. Vol. 275; no. 6; pp. 88-92

Heiderscheit, J.
For sale: Sulfur emissions
Independent Energy; Jan 31 1992
Description: The allowance trading market has started a slow march to maturity. Competitive developers should understand the risks and opportunities now presented. The marketplace for sulfur dioxide (SO_2) emissions allowances - the centerpiece of Title 4's acid rain reduction program - remains enigmatic 19 months after the Clean Air Act amendments of 1990 were passed. Yet it is increasingly clear that the emission allowance market will likely confound the gloom and doom of its doubters. The recently-announced $10 million dollar Wisconsin Power and Light allowance sales to Duquesne Light and the Tennessee Valley Authority are among the latest indications of momentum toward a stabilizing market. Vol. 226; pp. 9-12

Hemond, H.F.
Acid neutralizing capacity, alkalinity, and acid-base status of natural waters containing organic acids. Environmental Science and Technology; Oct 31 1990. Description: The terms acid neutralizing capacity (ANC) and alkalinity (Alk) are extensively employed in the characterization of natural waters, including soft circumneutral or acidic waters. However, in the presence of organic acids, ANC measurements are inconsistent with many conceptual definitions of ANC or Alk and do not provide an adequate characterization of the acid-base chemistry of water. Knowledge of Gran ANC and inorganic carbon concentrations does not by itself allow calculation of the pH of a water containing organic acids. Neither is measured ANC invariant upon changes in organic acid concentration. The result is a significant, but hidden, problem for policy makers and regulators, since ANC is considered a fundamental index of natural water acid-base status. ANC is the main output of many of the watershed simulation models now used in acid precipitation assessment programs, and considerations of present or expected ANCs apparently will play a major role in regulatory decisions. It is proposed to model such natural waters by (1) independently specifying the organic acid concentration and (2) adopting a definition of alkalinity that is invariant with changes of organic acid concentrations. Alkalinity, when so defined, can be both measurable and useful and possesses the conservative chemical properties commonly attributed to the term. A simple computational scheme, amenable to graphical presentation, is proposed to express the relationship between alkalinity, pH, organic acid concentration and inorganic carbon content.
Vol. 2410; pp. 1486-1489

Hendershot, W.H.; Warfvinge, P.; Sverdrup, H.U.; Courchesne, F.
Mobile anion concept - Time for a reappraisal. Journal of Environmental Quality; Jan 31 1991. Description: The mobile anion concept has been used to

support the argument that acid precipitation, containing elevated concentrations of nitrate and sulfate, is acidifying soils and surface waters. The authors believe that so much attention has been focused on the behavior of the strong acid anions that the effect of other important processes has, in some cases, been obscured. The emphasis, they believe, should be placed on processes that regulate H^+ in solution. Thus, they propose that the mobile anion concept (as an explanation of how acid precipitation degrades soils and surface waters) be replaced with descriptions of the mechanisms believed to control the movement of both anions and cations through ecosystems.
Vol. 203; pp. 505-509

Henrichs, R.; Cooper, L.I.
Law. Research Journal of the Water Pollution Control Federation; Jun 30 1990 Description: Administrative and judicial enforcement actions were a prominent aspect of the legal developments during 1989. Important regulatory and judicial developments also kept the interest of the regulatory community. However, 1989 will not be remembered for the passage of major federal environmental legislative initiatives or revisions
Vol. 624; pp. 320-338

Herlihy, A.T.; Kaufman, P.R.; Church, M.R.; Wigington, P.J. Jr.; Webb, J.R.; Sale, M.J.
The effects of acidic deposition on streams in the Appalachian Mountain and Piedmont region of the mid-Atlantic United States. Water Resources Research; Aug 31 1993. Description: Streams in the Appalachian Mountain area of the mid-Atlantic receive some of the largest acidic deposition loadings of any region of the US. A synthesis of the survey data from the mid-Appalachians yields a consistent picture of the acid base status of streams. Acidic streams, and streams with very low acid neutralizing capacity (ANC), are almost all located in small (<20 km^2), upland, forested catchments in areas of base-poor bedrock.
Vol. 298; pp. 2687-2703

Herlihy, A.T.; Kaufmann, P.R.; Mitch, M.E.
Stream chemistry in the eastern United States. 2. Current sources of acidity in acidic and low acid-neutralizing capacity streams. Water Resources Research; Apr 30 1991. Description: The authors examined anion composition in National Stream Survey (NSS) data in order to evaluate the most probably sources of current acidity in acidic and low acid-neutralizing capacity (ANC) streams in the eastern US. Acidic streams that had almost no organic influence (less than 10% of total anions) and sulfate and nitrate concentrations indicative of evaporative concentration of atmospheric deposition were classified as acidic due to acidic deposition. Vol. 274; pp. 629-642

Heslin, J.S.; Hobbs, B.F.
A probabilistic production costing analysis of SO_2 emissions reduction strategies for Ohio: Emissions, cost, and employment tradeoffs. Journal of the Air and Waste Management Association; Aug 31 1991. Description: A new approach for state- and utility-level analysis of the cost and regional economic impacts of strategies for reducing utility SO_2 emissions is summarized and applied to Ohio. The methodology is based upon probabilistic production costing and economic input-output analysis. It is an improvement over previous approaches because it: accurately models random outages of generating units, must-run constraints on unit output, and the distribution of power demands; and runs quickly on a microcomputer and yet considers the entire range of potential control strategies from a systems perspective. Vol. 418; pp. 956-966

Hessen, D.O.
Acidification of the HUMEX lake effects on epilimnetic pools and fluxes of carbon Environment International; Jan 31 1992 Description: The paper presents data on carbon budgets during the first year of acidification of one half of humic lake Skjervatjern. Dissolved organic carbon (DOC), mainly allochthonous humus, ranged from 2 to 12 mg C L^{-1}, with an average of 6 to 7 mg C for both the

acidified and the reference basin. The minimum values occurred during winter when ground temperature fell below zero. Average dissolved inorganic carbon (DIC) was near 500 [mu]g C L^{-1} in the surface layers, but increased to more than 2 mg close to the bottom. Epilimnetic particulate organic carbon (POC) was also close to 500 [mu]g C L^{-1}, of which more than 3/4 was detritus, while bacteria, phyto-, and macrozooplankton made up approximately 100, 50, and 20-40 [mu]g C L^{-1}, respectively. Vol. 186; pp. 649-657

Hicks, B.; McMillen, R.; Turner, R.S.; Holdren, G.R. Jr.; Strickland, T.C.
A national critical loads framework for atmospheric deposition effects assessment: III. Deposition characterization Environmental Management; Jan 31 1993 Description: Methods are discussed for describing patterns of current wet and dry deposition under various scenarios. It is proposed that total deposition data across an area of interest are the most relevant in the context of critical loads of acidic deposition, and that the total (i.e., wet plus dry) deposition will vary greatly with the location, the season, and the characteristics of individual subregions. Wet and dry deposition are proposed to differ in such fundamental ways that they must be considered separately.
Vol. 173; pp. 343-353

Hobara, Satoru.; Tokuchi, Naoko.; Ohte, Nobuhito.; Koba, Keisuke.; Katsuyama, Masanori.; Kim, Su-Jin.; Nakanishi, Asami.
Mechanism of nitrate loss from a forested catchment following a small-scale, natural disturbance. Canadian Journal of Forest Research; Aug 01 2001. Description: In Matsu-zawa catchment, central Japan, nitrate concentrations in stream water increased following a small-scale, natural disturbance involving an outbreak of pine wilt disease that affected ~25% of the forested catchment. To clarify nutrient dynamics in soils and their relationship with stream water nitrate, we investigated soil nitrogen dynamics and soil water chemistry in disturbed and undisturbed, water-unsaturated and -saturated plots. The highest values for nitrification rate, nitrate concentration in soil solution, and nitrate exported from the root zone were observed for the disturbed plot.
Vol. 31; no. 8; pp. 1326-1335

Hobbs, B.F.
Emissions dispatch under the underutilization provision of the 1990 U.S. Clean Air Act Amendments: Models and analysis. IEEE Transactions on Power Systems (Institute of Electrical and Electronics Engineers); Feb 28 1993. Description: The acid rain title of the new Clean Air Act will impact utility planning and operations in many ways. One important provision of the title will constrain the operation of coal-fired generating units that are subject to SO$_2$ limitations during Phase 1 of the Act (1995--99). Because only SO$_2$ emissions from those units will require emissions allowances during that time, utilities will be motivated to shift production to non-Phase 1 units whose SO$_2$ emissions are not limited until the year 2000.
Vol. 81; pp. 177-183

Hobbs, B.F.; Honious, J.C.; Bluestein, J.
Estimating the flexibility of utility resource plans: An application to natural gas cofiring for SO$_2$ control. IEEE Transactions on Power Systems (Institute of Electrical and Electronics Engineers); Feb 28 1994. Description: Utility planners must cope with large uncertainties concerning fuel prices, environmental laws, power demands, and the cost and availability of new resources. In this situation, flexibility is valuable. A flexible plan is one that enables the utility to quickly and inexpensively change the system's configuration or operation in response to varying market and regulatory conditions. Vol. 91; pp. 167-173

Hobbs, B.F.; Wilson, A.F.
Most value planning: Estimating the net benefits of electric utility resource plans Energy Sources; Jan 31 1994. Description: Most US utility regulatory commissions require that electric utilities minimize cost

when comparing supply-side and demand-side resources. However, utilities should also consider the effects of resource plans upon the benefits, or value", that electricity consumers receive. A method for extending the traditional least cost" objective to include value changes in electric utility resource planning is presented. Vol. 163; pp. 451-477

Hoelldampf, B.; Barker, A.V.
Effects of ammonium on elemental nutrition of red spruce and indicator plants grown in acid soil. Communications in Soil Science and Plant Analysis; Jan 31 1993. Description: Decline of high elevation red spruce forests in the northeastern United States has been related to acid rain, particularly with respect to the deposition of nitrogenous materials. Ca and Mg deficiencies may be induced by input of air- borne nitrogenous nutrients into the forest ecosystem. This research investigated the effects of N nutrition on mineral nutrition of red spruce and radish, as an indicator plant, grown in acid forest soil. Vol. 241516; pp. 1945-1957

Hoelldampf, B.; Barker, A.V.; Smith, G.
Mineral element composition of declining and healthy stands of red spruce in western Massachusetts. Communications in Soil Science and Plant Analysis; Jan 31 1993. Description: The appearance of yellowing and loss of needles in conifers of the high-elevation forest of the NE US has been attributed to acidic fog of clouds. Long-term effects of acid precipitation are reported to injury plant canopies and alter soil chemistry, including changes in availability of nutrients.
Vol. 241516; pp. 1937-1944

Hofmann, D.J.
Increase in the stratospheric background sulfuric acid aerosol mass in the past 10 years. Science (Washington, D.C.); May 25 1990. Description: Data obtained from measurements of the stratospheric aerosol at Laramie, Wyoming (41° N), indicate that the background or nonvolcanic stratospheric sulfuric acid aerosol mass at northern mid-latitudes has increased by about 5 ± 2% per year during the past 10 years. Whether this increase is natural or anthropogenic could not be determined at this time because of inadequate information on sulfur sources, in particular, carbonyl sulfide, which is thought to be the dominant nonvolcanic source of stratospheric sulfuric acid vapor. An increase in stratospheric sulfate levels has important climatic implications as well as heterogeneous chemical effects that may alter the concentration of stratospheric ozone. Vol. 2484958; pp. 996-1000

Hoisve, R.A.; Blanc, F.C.
Effect of pH on leaf decomposition in low alkalinity lakes. Journal of Environmental Engineering (New York); Jan 31 1993. Description: Two microcosm experiments were conducted to study the effects of low pH on leaf decomposition in low-alkalinity lakes. In the first experiment, a mix of oak and birch litter was incubated for six months in water from Little Rock Lake, Wisconsin, at pH 5.0 and 6.5. In the second experiment, maple litter was incubated for three months in water from Round Pond, Massachusetts, at pH 4.0, and in water from Walden Pond, Massachusetts, at pH 6.0. In both experiments, litter weight loss was less at the lower pH. However, the initial carbon loss from maple litter at pH 4.0 was faster than at pH 6.0 in the second experiment. Vol. 1191; pp. 56-71

Holzman, D.
Species extinction mires ecosystem Insight; Mar 26 1990. Description: Extinction is normal in the evolution of life, but amphibians, insects, birds and mammals are vanishing at an alarming pace. While habitat destruction, overexploitation and pollution are among the main causes, some disappearances cannot be explained. The extinction problem among amphibians mirrors the general, worldwide phenomenon. A synergism of insults may be responsible. Chance events such as a dry year might occasionally clean out a pond. But a larger lake nearby would replenish it. Now acid pollution adds to the ponds' burden while stocking of amphibian-eating sport fish in

the lake - which happens even in natural parks - would destroy the source of replenishment. Some fear that extinctions ultimately could destroy nature's fabric. Vol. 613; pp. 52-53

Hordijk, L.
Use of the RAINS model in acid rain negotiations in Europe. Environmental Science and Technology; Apr 30 1991. Description: The use of models in international negotiations on environmental problems for which no compulsory action can be imposed is a recent trend. In the past, international agreements have been reached without any model being used. For example, the first step in reducing acid rain in Europe and North America was made in 1985 without using an integrated model. Neither was a model used to establish the Vienna Convention on Protection of the Ozone Layer (1986). Vol. 254; pp. 596-603

Hoske, M.T.
Phase I compliance plans emphasize flexibility. Electric Light and Power (Boston); Aug 31 1993. Description: Most utilities have emphasized flexibility in strategies for complying with Phase I of the Clean Air Act Amendments (CAAA) of 1990. Nearly half of the compliance methods involve fuel blending. Using allowances to comply also turned out more popular than many anticipated. Both methods give utilities flexibility to change strategies should resolving uncertainties favor one compliance method over another. Vol. 718; pp. 8, 11

Hough, A.M.
Development of a two-dimensional global tropospheric model: Model chemistry Journal of Geophysical Research; Apr 20 1991. Description: A latitudinally averaged two-dimensional model has been used to study the distributions, budgets, and trends of trace gases in the atmosphere from pole to pole and from the surface to 24 km. The chemical mechanism used contains 56 chemical species, including 12 hydrocarbons and 125 chemical and photochemical reactions, as well as wet removal processes and dry deposition. Apart from the stratospheric sources of ozone and nitrogen oxides the model chemistry is driven completely by the time-dependent photolytic processes and the emission of 17 chemical species distributed according to 10 different source categories. Vol. 964; pp. 7325-7362

Houle, Daniel.; Paquin, Raynald.; Camiré, Claude.; Ouimet, Rock.; Duchesne, Louis.
Response of the Lake Clair Watershed (Duchesnay, Quebec) to changes in precipitation chemistry (1988-1994) Canadian Journal of Forest Research; Nov 01 1997. Description: The chemistry of precipitation and of the lake's outlet (1988-1994) were measured at the Lake Clair Watershed (226 ha, 46°57 prime N, 71°40 prime W, 270-390 m above sea level), which is located 50 km northwest of Québec City, Quebec, Canada. In wet precipitation, concentrations of SO_4, Ca, and Na decreased from 1988 to 1994 whereas pH increased. In bulk precipitation, only Ca and Na decreased. The lake's outlet SO_4 concentration decreased from 1988 to 1994, suggesting that the catchment rapidly responded to the changes in precipitation although a net SO_4-S export was observed each year between 1988 and 1994.
Vol. 27; no. 11; pp. 1813-1821

Houpis, J.L.J.; Costella, M.P.; Cowles, S.
A branch exposure chamber for fumigating ponderosa pine to atmospheric pollution Journal of Environmental Quality; Jan 31 1991. Description: This paper describes the design, construction, and testing of an alternative tool to whole-tree enclosures for measuring pollution response in mature woody tissue. The chamber is a new design, though not a new concept, and is referred to as a branch exposure chamber. Designed primarily for ozone and acid precipitation exposures (and used additionally for CO_2 measurements), the branch exposure chamber incorporates four major parts: support structure, fan-air supply unit, charcoal filter unit, and exposure chamber. The exposure chamber is a 1.5-m long by 0.7-m diam. cylinder.

The chamber is constructed of Teflon tape. Three zones in the chamber affect exposure of the experimental tissue: an initial buffer region for mixing, a main exposure region, and an exhaust frustrum. Aerodynamic testing of the chamber-mixing characteristics show that mixing is uniform and complete within the main exposure region. Thermal buildup within the chamber was a maximum of 3C under a wide range of ambient meteorological conditions. Based on current field trials of the chamber, material deterioration due to environmental variables (e.g., ultraviolet radiation, heat oxidants), is not expected to affect operation of the chamber for 24 mo. The BEC is inexpensive to build and operate, and represents a viable alternative to a whole-tree chamber.
Vol. 202; pp. 467-474

Hov, Oe.
Some reasons for changes in acid deposition during the last decades
International Journal of Energy-Environment-Economics; Jan 31 1992
Description: Published data show that the atmospheric concentration of sulfate over Europe has probably increased by a factor of 3 since 1900, while the nitrate levels have risen dramatically over the last 30--40 years with a doubling from 1955 to 1985. The meteorological variability can on an annual basis give rise to as much as a 20% change in the average concentration in SO_2 even if the emissions are kept constant. This makes it difficult to detect trends due to emission changes in the atmospheric concentration of e.g. SO_2.
Vol. 23; pp. 177-186

Huang, Wenxiong; Hobbs, B.F.
Optimal SO_2 compliance planning using probabilistic production costing and generalized benders decomposition
IEEE Transactions on Power Systems (Institute of Electrical and Electronics Engineers); Feb 28 1994
Description: In 1990, the US Congress a new Clean Air Act which contains provisions to control sulfur dioxide (SO_2, a primary cause of acid rain) emitted from electric generation plants in the US. Under this Act, electric utilities will be able to choose from a wide range of SO_2 emissions control measures. This paper presents a comprehensive emissions control model which can systematically examine all available emissions control options and construction optimal compliance plan. The model is a nonlinear integer program that uses probabilistic production costing to simulate system generation. Vol. 91; pp. 174-180

Huang, Y.J.; Wang, H.P.
Reduction of NO with CH_4 effected by copper oxide clusters in the channels of ZSM-5. Journal of Physical Chemistry A: Molecules, Spectroscopy, Kinetics, Environment, amp General Theory; Aug 19 1999. Description: Nitrogen oxides (NO_x) are the main air pollutants that cause photochemical smog formation, acid rain, and general atmospheric visibility degradation. Worldwide, over 3×10^7 tons of nitrogen oxides are vented to the atmosphere each year. Reduction of NO_x emission has become one of the major problems to be solved for environmental protection. Vol. 103; no. 33; pp. 6514-6516

Hubbard, H.M.
The real cost of energy
Scientific American; Apr 30 1991
Description: Gas prices only seem high. When you say fillerup, you pay but a fraction of the actual cost. Not included are the tens of billions (close to $50 for each barrel of oil) the military spends annually to protect oil fields in the Persian Gulf. Then tack on the hidden costs of environmental degradation, health effects, lost employment, government subsidies and more. Sooner or later, the public pays the entire price. Bringing market prices in line with energy's hidden burdens will be one of the great challenges of the coming decades. The author describes these hidden costs and makes estimates of them.
Vol. 2644; pp. 36-40,42

Hudson, J.G.; Svensson, G.
Cloud microphysical relationships in California marine stratus
Journal of Applied Meteorology; Dec 31

1995. Description: Cloud microphysical measurements off the southern California coast are presented and compared with in situ airborne measurements of cloud condensation nuclei (CCN) spectra. Large-scale variations in cloud droplet concentrations were due to CCN variations, some medium-scale variations may be a result of the conversion of droplets to drops by coalescence, while small-scale variations were due to different proportions of the CCN spectra being activated because of variations in updraft velocity at cloud base.
Vol. 34; no. 12; pp. 2655-2666

Hughes, R.N.; Cox, R.M.
Acidic fog and temperature effects on stigmatic receptivity in two birch species Journal of Environmental Quality; Jul 31 1994. Description: Factorial assays were performed to determine the effects of simulated acid fog (SAF) and temperature on stigmatic receptivity in two birch species. Excised reproductive branches were sampled from representative individuals of mountain paper birch (Betula cordifolia Regel.) and paper birch (Betula papyrifera Marsh.) in populations adjacent to the Bay of Fundy, New Brunswick, Canada. Since 1979 these trees have exhibited branch dieback in association with abnormal foliar browning symptoms. This browning has been linked with acidity and nitrate deposited by fog, which is frequent in the area. In general, experimental results indicated that pollen germination increased with temperature, but pH effects were less obvious.
Vol. 23; no. 4; pp. 686-692

Hughes, R.N.; Cox, R.M.
In vitro pollen responses of two birch species to acidity and temperature Journal of Environmental Quality; Oct 31 1993. Description: Paper birch (Betula papyrifera Marsh.) and mountain paper birch (Betula cordifolia Regel) near the Bay of Fundy coast frequently intercept acidic advection marine fogs. Chemical deposition by these fogs is thought to be a factor contributing to the observed foliar browning symptoms associated with a marked deterioration of these trees in the area. In vitro experiments were performed to test whether pollen germination in these two birch species would be affected by acidity at levels routinely found in the fog.
Vol. 22; no. 4; pp. 799-804

Huret, N; Chaumerliac, N; Isaka, H; Nickerson, E.C.
Influence of different microphysical schemes on the prediction of dissolution of nonreactive gases by cloud droplets and raindrops. Journal of Applied Meteorology; Sep 30 1994. Description: Three microphysical formulations are closely compared to evaluate their impact upon gas scavenging and wet deposition processes. They range from a classical bulk approach to a fully spectral representation, including an intermediate semispectral parameterization. Detailed comparisons among the microphysical rates provided by these three parameterizations are performed with special emphasis on evaporation rate calculations. This comparative study is carried out in the context of a mountain wave simulation.
Vol. 33; no. 9; pp. 1096-1109

Hutchinson, Thomas.C.; Watmough, Shaun.A.; Sager, Eric.P.; Karagatzides, Jim.D.
Effects of excess nitrogen deposition and soil acidification experimental study Canadian Journal of Forest Research; Feb 01 1998. Description: The impact of an acidifying fertilizer on litter decomposition, root mycorrhizae, and soil and tree chemistry was assessed in two hardwood forests in central Ontario, Canada. Soil beneath mature sugar maple (*Acer saccharum* Marsh.) trees was treated with $(NH_4)_2SO_4$ granules at application rates of 0, 250, 500, and 1000 kg/ha in May of each year between 1993 and 1994 at Dorset and between 1993 and 1995 at Loring. The fertilizer treatments did not cause visual symptoms of forest decline. At Dorset, SO_4 and cation concentrations in soil leachate increased, but no difference in soil pH between treatments was found
Vol. 28; no. 2; pp. 299-310

Igolkina, E.D.
> Variations in the acid-alkali balance of natural waters and some aspects of establishing ecological standards. Water Resources; Jul 31 1995. Description: Conditions of hydrogen-ion supply and their effect on water ecosystems are discussed, and the concept of ecological pH standard for different water bodies is formulated. The idea of ecological reserve of aquatic ecosystems with respect to the pH factor is suggested. Establishing ecological standards by analogy is discussed. Vol. 22; no. 4; pp. 380-38

Irwin, B.
> NO_x reduction techniques. American Ceramic Society Bulletin; Oct 31 1995 Description: NO_x is the combination of nitrogen and oxygen that contributes to acid rain and smog problems. It is formed in high-temperature flames and processes, including kilns and furnaces. Reducing this pollutant is a universal concern, and its regulation will be increased as individual states devise their implementation plans for complying with the US Clean Air Act. Vol. 74; no. 10; pp. 81-84

Ishihara, Shigehis; Furutsuka, Takeshi
> Removal of NO_x or its conversion into harmless gases by charcoals and composites of metal oxides. Preprints of Papers, American Chemical Society, Division of Fuel Chemistry; Jan 31 1996. Description: In recent years, much attention has been devoted to environmental problems such as acid rain, photochemical smog and water pollution. In particular, NO_x emissions from factories, auto mobiles, etc. in urban areas have become worse. To solve these problems on environmental pollution on a global scale, the use of activated charcoal to reduce air pollutants is increasing. However, the capability of wood-based charcoal materials is not yet fully known. The removal of NO_x or its conversion into harmless gases such as N_2 should be described. In this study, the adsorption of NO over wood charcoal or metal oxide-dispersed wood charcoal was investigated. Vol. 41; no. 1; pp. 289-292

Ishizuka, T.; Kabashima, H.; Yamaguchi, T.; Tanabe, K.; Hattori, H.
> Initial step of flue gas desulfurization - an IR study of the reaction of SO_2 with NO_x on CaO. Environmental Science and Technology; Jul 01 2000. Description: Flue gas desulfurization contributes much to the prevention of acid rain. In a dry-type flue gas desulfurization, promotive effect of NO on SO_2 absorption is observed. To elucidate the reaction mechanisms for absorption of SO_2 by a calcium compound in a dry-type flue gas desulfurization, the states of adsorbed SO_2, NO and NO_2 on CaO and the reactivity of the adsorbed SO_2 with NO, NO_2 and O_2 were studied by IR and temperature-programmed desorption (TPD). Sulfur dioxide (SO_2) and NO_2 were adsorbed on CaO mainly in the form of sulfite ion (SO_3^{2-}) and nitrato (NO_3) complex respectively.
> Vol. 34; no. 13; pp. 2799-2803

Itoh, Mik; Nishihara, Hirosh; Aramaki, Kunitsugu
> The protection ability of 11-mercapto-1-undecanol self-assembled monolayer modified with alkyltrichlorosilanes against corrosion of copper. Journal of the Electrochemical Society; Jun 30 1995. Description: Closely packed polymer films of monolayers adsorbed on the copper surface were prepared by chemical modification of a 11-mercapto-1-undecanol $HO(CH_2)_{11}SH$ (MUO) self-assembled monolayer with alkyltrichlorosilanes $C_nH_{2n+1}SiCl_3$ (C_nTCS) and water. The maximum protection efficiency, 94.7% for the MUO monolayer modified with $C_{18}TCS$ in a dilute solution against copper corrosion in aerated 0.5M Na_2SO_4 was obtained by polarization and impedance measurements. A Fourier transform infrared reflection absorption spectrum of the surface covered with the monolayer showed the absence of hydroxyl groups in the layer, implying the formation of a regularly arranged polymer monolayer film on the surface.
> Vol. 142; no. 6; pp. 1839-1846

Iwamoto, Masakazu; Yahiro, Hidenori; Tanda, Kenji; Mizuno, Noritaka; Mine, Yosihiro;

Kagawa, Shuichi
Removal of nitrogen monoxide through a novel catalytic process. 1. Decomposition on excessively copper ion exchanged ZSM-5 zeolites. Journal of Physical Chemistry; May 02 1991. Description: Repeated ion exchange of the ZSM-5 zeolite using aqueous copper(II) acetate solution was found to bring about excess loading of copper ions above an exchange level of 100%. The high activity of the resulting catalyst for NO decomposition was consistent for at least 30 h even at short contact time and low NO pressure. The number of copper ions that can adsorb NO molecules has been determined by a temperature-programmed desorption technique combined with IR measurement; 94% of Cu^{2+} ions in ZSM-5 were active for the adsorption. Vol. 959; pp. 3727-3730

J., Yuegang Zuo, Hoigne,
Evidence for photochemical formation of H_2O_2 and oxidation of SO_2 in authentic fog water. Science (Washington, D.C.); Apr 02 1993. Description: When samples of rain and fog water were exposed to ultraviolet and visible light, reactive transients such as hydrogen peroxide were formed and dissolved organic matter and sulfur dioxide were depleted. These results, in conjunction with those from previous studies, imply that dissolved organic compounds and transition metals such as iron ions are involved in the photochemical formation of hydrogen peroxide and other photooxidants in atmospheric waters. Vol. 2605104; pp. 71-73

J.E., de Steiguer,; Pye, J.M.; Love, C.S.
Air pollution damage to U.S. forests Journal of Forestry; Aug 31 1990 Description: A survey was made of the perceptions and estimates by forestry and air pollution experts, using a Delphi procedure with three sequentially mailed questionnaires. Five pollutants: ozone, sulfuric acid, nitric acid, sulfur dioxide and nitrogen oxides; and seven forest ecosystems in the us were included. Of the five pollutants at ambient levels, ozone is perceived as the greatest threat to forest growth. In contrast acid deposition is thought to cause growth reductions only in Appalachian high-elevation spruce-fir forests. Vol. 888; pp. 17-22

Jager, H.I.; Sale, M.J.; Schmoyer, R.L.
Cokriging to assess regional stream quality in the Southern Blue Ridge Province Water Resources Research; Jul 31 1990 Description: Cokriging is used to predict stream chemistry at unsampled locations with the use of spatial and intervariable correlation. The technique is used in this study to predict the acid neutralizing capacity (ANC) of streams in the Southern Blue Ridge Province (SBRP). ANC measurements between pairs of streams surveyed in this region were found to be spatially correlated over distances up to around 40 km. Predictions were improved by including elevation in the analysis to represent the combined influence of elevational gradients in climate, geology, soils, hydrology, and vegetation on stream ANC. The cokriging analysis identified specific stream reaches predicted to be most sensitive to acidification and located areas of high uncertainty. Stream ANC levels below 50 {mu}eq/L were predicted for one-fifth of the upper nodes associated with digitized headwater reaches in the SBRP. The majority of these were lcoated in the higher elevations of the Great Smoky Mountains National Park, in the vicinity of Mount Mitchell, and in the Blue Ridge Mountains in southern North Carolina. Vol. 267; pp. 1401-1412

Jayanty, R.K.M.; Gay, B.W. Jr.
Measurement of toxic and related air pollutants. Journal of the Air and Waste Management Association; Dec 31 1990 Description: A joint conference for the fifth straight year cosponsored by the Air and Waste Management Association's EM-3, EM-4, and ITF-2 technical committees, and the Atmospheric Research and Exposure Assessment Laboratory (AREAL) of the US Environmental Protection Agency, was held in Raleigh, North Carolina, May 1-4, 1990. The technical program consisted of 187 presentations, held in 20 technical sessions, on recent advances in the measurement and

monitoring of toxic and related pollutants found in ambient and source atmospheres Vol. 4012; pp. 1631-1637

Jayne, J.T.; Gardner, J.A.; Davidovits, P.; Worsnop, D.R.; Zahniser, M.S.; Kolb, C.E.
The effect of H_2O_2 content on the uptake of SO_2 (g) by aqueous droplets
Journal of Geophysical Research; Nov 20 1990. Description: The effect of H_2O_2 content on the uptake of SO_2 by aqueous droplets has been measured. For pH < 4, SO_2 gas uptake increased when the aqueous solution was doped with (H_2O_2) up to 2 M. The dependence of the rate of uptake on (H_2O_2) is consistent with the reaction rate of H_2O_2 with S(IV) measured in bulk aqueous solutions indicating that the droplet surface does not play a significant role in the H_2O_2- S(IV) reaction.
Vol. 9512; pp. 20,559-20,563

Jeffries, D.S.; Lam, D.CL.; Wong, I.; Moran, M.D.
Assessment of changes in lake pH in southeastern Canada arising from present levels and expected reductions in acidic deposition. Canadian Journal of Fisheries and Aquatic Sciences; Feb 01 2000. Description: An integrated acid rain assessment model was used to estimate pH for six clusters of lakes in southeastern Canada and scenarios of sulphate deposition that reflect the situation (*a*) before implementation of the SO_2 emission controls required by the Canada/U.S. Air Quality Agreement, (*b*) after implementation of Canadian controls, and (*c*) after implementation of Canadian and U.S. controls. Modelled lake pHs were always less than their estimated original values. To assess the ecological significance of the pH reduction, scenario "damage" was quantified as the percentage of cluster lakes having pH < 6, a threshold criterion sufficient to protect most aquatic biotave little effect in Atlantic Canada.
Vol. 57; no. 2; pp. 40-49

Johansson, Per.; Gustafsson, Lena.
Red-listed and indicator lichens in woodland key habitats and production forests in Sweden. Canadian Journal of Forest Research; Sep 01 2001. Description: There are ca. 70 000 "woodland key habitats" (WKHs) in Sweden that according to definition should contain red-listed species, but their species content is seldom known. Indicator species are used as one tool to identify the WKHs. In two areas in southern Sweden red-listed and indicator lichen species were surveyed in line transects in a total of 25 WKHs (45 ha) and, for comparison, in 74 ha of surrounding production forest.
Vol. 31; no. 9; pp. 1617-1628

Johnson, B.D.
Nitrate and pH monitoring of rainfall and its effect on west Georgia streams
Georgia Journal of Science; Jan 31 1995 Description: Beginning July 1, 1994, rain was collected from a site in Carrollton, Georgia, and a site in Cedartown, Georgia, and measured for nitrate and pH. In conjunction with the rainwater analysis, water samples were collected from five sites in the Piedmont Province along the Little Tallapoosa River near Villa Rica, Georgia. Water samples were also taken from twelve sites along Cedar Creek, near Cedartown, Georgia, Cedar Creek flows across limestone, dolostones, and shales of the Valley and Ridge Province. These samples were also measured for nitrate and pH. Vol. 53; no. 1; pp. 34-35

Johnson, Chris.E.; Romanowicz, Rachel.B.; Siccama, Thomas.G.
Conservation of exchangeable cations after clear-cutting of a northern hardwood forest Canadian Journal of Forest Research; Jun 01 1997. Description: Clear-cut logging of northern hardwoods disrupts forest nutrient cycling, often enhancing export of mineral nutrients (Ca, Mg, K) in drainage waters. When these leaching losses are added to mineral nutrient export in harvested wood, the total can be a significant fraction of the standing stocks of some ecosystems. Thus, changes in soil chemical properties after logging are important in determining the long-term implications of harvesting on nutrient availability and site fertility.
Vol. 27; no. 6; pp. 859-868

Johnson, D.W; Susfalk, R.B.; Brewer, P.F.
Simulated responses of red spruce forest soils to reduced sulfur and nitrogen deposition. Journal of Environmental Quality; Nov 30 1996. Description: Implications of reducing S and N deposition on red spruce forests of the southern Appalachians are explored using the Nutrient Cycling Model (NuCM). We hypothesized that reducing deposition would cause (i) large reductions in soil solution NO_3^-, SO_4^{2-}, Al and Ca/Al ratios, but (II) small changes in exchangeable base cation reserves. Hypothesis (I) was supported in part: simulated reductions in atmospheric deposition had substantial and nearly immediate effects upon soil solution mineral acid anions, Ca^{2+}, and Al^{3+} concentrations. Ca/Al molar ratios were much less sensitive to changes in deposition and soil solution ionic strength than either Ca^{2+} or Al^{3+} separately. Hypothesis (II) was not supported: although the increases in base saturation and exchangeable cation pools were small relative to cation exchange, they were large relative to initial exchangeable base cation pools. Vol. 25; no. 6; pp. 1300-1309

Johnson, G.L.
Searchers for a new energy source: Tesla, Moray, and Bearden
IEEE Power Engineering Review (Institute of Electrical and Electronics Engineers); Jan 31 1992. Description: Tesla, Moray, Bearden, and others have claimed the existence of another source of energy besides those presently in use. Like sun and wind, this source is available without regard to political boundaries. If true, the development of this energy source would be one of the most important events of the century. It seems that every time mankind reaches a limit of growth due to exhaustion of inexpensive energy supplies, another energy sources is discovered and developed. England had essentially depleted its resources of timber when the technology to mine and burn coal was developed, for example. After coal, technologies for oil, gas, hydro, nuclear fission, wind, photovoltaic, etc. were developed. With each new development, the world was able to support a greater population at a higher standard of living than before. Vol. 121; pp. 20-22

Joskow, P.L.
Weighing environmental externalities: Let's do it right. Electricity Journal; May 31 1992. Description: Should we as a society adopt policies to internalize external environmental costs Of course we should. But we should do it correctly. State public utility commissions (PUCs) that are using numerical externality adders' reflecting global and regional environmental impacts in the resource planning and selection process are doing it wrong. Vol. 54; pp. 53-67

Joslin, J.D.; Kelly, J.M.; H., Van Miegroet,
Soil chemistry and nutrition of North American spruce-fir stands: Evidence of recent change. Journal of Environmental Quality; Jan 31 1992. Description: One set of hypotheses offered to explain the decline of red spruce (Picea rubens Sarg.) in eastern North America focuses on the effect of acidic deposition on soil chemistry changes that may affect nutrient availability and root function. Long-term soils data suggests that soil acidification has occurred in some spruce stands over the past 50 yr, with plant uptake and cation leaching both contributing to the loss of cations. Vol. 211; pp. 12-30

Kaiser, J
Acid rains' dirty business: Stealing minerals from soil. Science (Washington, D.C.); Apr 12 1996. Description: This article describes the hidden environmental effects of acid rain - leaching of base mineral ions from the soil, often changing soil chemistry dramatically. The primary information comes from Ecosystem studies at Hubbard Brook of Likens and Buso. The article also discusses both other opinions and possible solutions.
Vol. 272; no. 5259; pp. 198

Kaplan, D
Wind power finding its competitive edge Energy Daily; Aug 18 1993
Description: When interviewing the head

of the windpower association, one expects to hear a barrage of global warming, acid rain and other pollution horror stories, followed by a call for an expensive federal effort to replace fossil fuels with wind power. But not from Randy Swisher, president of the American Wind Energy Association. This article describes the technological advances made in wind energy during the last decade, and its cost competitiveness with conventional fossil fuels. Vol. 21; no. 158; pp. 1, 4

Kaufman, Y.J.; Chou, M.D.
Model simulations of the competing climatic effects of SO_2 and CO_2
Journal of Climate; Jul 31 1993
Description: Sulfur dioxide-derived cloud condensation nuclei are expected to enhance the planetary albedo, thereby cooling the planet. This effect might counteract the global warming expected from enhanced greenhouse gases. A detailed treatment of the relationship between fossil fuel burning and the SO_2 effect on cloud albedo is implemented in a two-dimensional model for assessing the climate impact. Some general conclusions can be reached. Vol. 67; pp. 1241-1252

Kaufmann, P.R.; Herlihy, A.T.; Mitch, M.E.; Messer, J.J.; Overton, W.S.
Stream chemistry in the eastern United States. 1. Synoptic survey design, acid-base status, and regional patterns
Water Resources Research; Apr 30 1991
Description: To assess the regional acid-base status of streams in the mid-Atlantic and southern US, spring base flow chemistry was surveyed in a probability sample of 500 stream reaches representing a population of 64,300 reaches (224,000 km). Approximately half of the streams had acid-neutralizing capacity (ANC) {le} 200 {mu}eq L^{-1}. Acidic (ANC {le} 0) streams were located in the highlands of the Mid-Atlantic region (southern New York to southern Virginia, 2,330 km), in coastal lowlands of the Mid-Atlantic (2,600 km), and in Florida (462 km). Vol. 274; pp. 611-627

Keller, W.; Dixit, S.S.; Heneberry, J.
Calcium declines in northeastern Ontario lakes. Canadian Journal of Fisheries and Aquatic Sciences; Oct 01 2001.
Description: Thousands of lakes in northeastern Ontario, Canada, have been acidified by sulphur deposition associated with emissions from the Sudbury area metal smelters. However, water quality improvements including increased pH and reduced sulphate concentrations have followed large reductions in Sudbury emissions that were implemented, beginning in the 1970s. Substantial decreases in Ca concentrations accompanied these other changes in lakewater chemistry. Monitoring of 38 lakes 20–128 km from Sudbury showed declines in Ca concentrations, averaging 2.7 µeq·L^{-1}·year^{-1}, over the period 1981–1999. Vol. 58; no. 10; pp. 2011-2020

Keller, W.; Yan, N.D.; Holtze, K.E.; Pitblado, J.R.
Inferred effects of lake acidification on Daphnia galeata mendotae
Environmental Science and Technology; Aug 31 1990. Description: Large numbers of Canadian Shield lakes have been acidified by the atmospheric deposition of anthropogenic sulfur. Biological damage attributable to acidification occurs at all levels of aquatic food webs; however, documentation of this damage has largely been confined to areas near large point sources of air pollutants, to small numbers of study lakes, or to experimentally acidified lakes. Demonstrations of widespread biological effects of acidification have been greatly hampered by the general absence of observations of the occurrence or abundance of important, ubiquitous species in large numbers of lakes ranging widely in acidity, coupled with laboratory determinations of lethal acid thresholds for these species. In consequence, it has been necessary to estimate rather than to document the regional extent of biological damage in North America. In this report the authors couple determination of the lethal acid threshold of Daphnia galeata mendotae Birge, a large, ubiquitous, planktonic

crustacean, with results of extensive lake surveys, to examine if the acidification of lakes in Ontario has resulted in widespread losses of this important member of the zooplankton. Vol. 248; pp. 1259-1261

Kern, E; Polansky, A
How many rooftop PV systems does it take to save the planet? Solar Industry Journal; Jan 31 1993. Description: This paper describes a program being managed by the United States Environmental Protection Agency (EPA) at the Air and Energy Engineering Research Laboratory which houses EPA's air-quality research and development activities. EPA is managing a program to collect field data of the energy savings and air pollution reductions achieved by installing rooftop photovoltaic systems (PV) on residential and commercial buildings. Emission offsets will be measured for the acid-rain, greenhouse gases. The goal is to demonstrate the effectiveness of solar technologies as a pollution-mitigating energy replacement for fossil fuels.
Vol. 4; no. 4; pp. 18-28

Kieber, R.J; Rhines, M.F; Willey, J.D; Avery, G.B. Jr.
Nitrate variability in coastal North Carolina rainwater and its impact on the nitrogen cycle in rain. Environmental Science and Technology; Feb 01 1999. Description: The concentration range for nitrite (NO_2^-) in 115 rain samples collected in Wilmington, NC, from June 1996 through February 1998 was 0.022--0.603 {micro}M. Nitrite concentrations did not correlate with precipitation volume, suggesting a continuous supply of nitrite during rain events possibly by slow scavenging of gas-phase material such as $HONO(g)$ or $NO_2(g)$ or in-cloud oxidation of other reduced forms of nitrogen. Nitrite levels exhibited no seasonal oscillations, which is in contrast to other rainwater parameters at this site such as pH, nitrate, non-seasalt sulfate (NSS) and ammonium.
Vol. 33; no. 3; pp. 373-377

Kim, A.
Utilization of coal combustion by-products: determining the environmental safety. Energeia; Jul 01 2001. Description: In the USA there is economic and environmental impetus to increase the utilization of coal combustion by-products (CCBS). CCBs contain small mounts of trace elements. The potential release of trace elements from CCBs by natural liquids is a factor in estimating the environmental risk of beneficial uses such as bulk fill and mine remediation
Vol. 12; no. 2; pp. 1-3

Kim, D.S.; Aneja, V.P.
Microphysical effects on cloud water acidity: A case study in a non precipitating cloud event observed at Mt. Mitchell, North Carolina. Air and Waste; Oct 31 1992. Description: There is increasing recognition that deposition of cloud water may significantly contribute to the decline of forest. In this paper, chemical and microphysical data obtained at Mt. Mitchell, North Carolina during one orographic, non- precipitating cloud event are examined to investigate the relationship between temporal variation of cloud acidity and cloud microphysics. The cloud acidity was substantially higher than in precipitating cloud events. Sulfate, nitrate, ammounium and hydrogen ions were the major constituents.
Vol. 4210; pp. 1345-1349

Kim, Y.J.; Boatman, J.F.
The collection efficiency of a modified Mohnen slotted-rod cloud-water collector in summer clouds. Journal of Atmospheric and Oceanic Technology; Feb 28 1992 Description: A modified Mohnen slotted-rod collector was used to collect cloud-water samples in summer clouds over the northeastern United States. Cloud-droplet-size distributions were measured with a forward-scattering spectrometer probe (FSSP) mounted on the NOAA King Air research aircraft. Cloud-droplet-volume distributions and liquid water content were determined for each cloud-water sample through analyses of the FSSP data.
Vol. 9; pp. 35-41

King, G.A.
> Corrosivity mapping -- A novel tool for materials selection and asset management Materials Performance; Jan 31 1995 Description: Studies worldwide have shown that the overall cost of corrosion amounts to at least 2 to 3% of the gross national product and that 20 to 25% of that cost could be avoided by using current corrosion control technology. Atmospheric corrosion is the major contributor to this cost. Having the tools and knowledge to initially specify the proper materials or corrosion protection for a structure or identify the optimum intervals for maintenance would generate considerable savings. Vol. 34; no. 1; pp. 6-9

Kirkwood, D.E.; Nesbitt, H.W.
> Formation and evolution of soils from an acidified watershed: Plastic Lake, Ontario, Canada. Geochimica et Cosmochimica Acta; May 31 1991. Description: The Plastic Lake watershed contains podzols developed on glacial tills deposited 12,000 years ago. Present-day, cationic fluxes from the soils are greater by a factor of 2 than long-term fluxes averaged over the age of the tills. The high rates of present-day chemical weathering may be a result of increased input of anthropogenic acids into the Plastic Lake watershed. Time-averaged proportions of cations leached from the soils are strikingly different from the proportions of cations now being leached, indicating that the character of chemical weathering has changed over time. Vol. 555; pp. 1295-1308

Kissam, A.D
> Pollution control for cash. Independent Energy; Jan 31 1995. Description: Significant amounts of money are now trading hands in the market for pollution allowance credits. Operators or developers of fossil fuel power plants in the United States will want to be aware of the financial implications and how it can affect the bottom line. It will be increasingly important to project participants that the emission credit issue is covered in the plan. The credits may be so cheap today that years of requirements can be purchased for less than avoided cost of the capital equipment for pollution control. Vol. 25; no. 1; pp. 52-54

Kline, T.R; Porter, J.M; Hannapel, J.S; Panzik, S.
> Energy resources law: Update on environmental and health and safety regulatory issues. Tort and Insurance Law Journal; Jan 31 1993. Description: This article provides an update on several environmental and health and safety issues that impact the development, management, and use of energy resources. Specifically, regulatory developments involving waste management activities under the Resource Conservation and Recovery Act (RCRA), including threshold issues such as the definition of waste under RCRA (i.e., the mixture and derived-from rules), are included in this article. In addition, new regulations on used oil recycling management standards and land disposal restriction for hazardous debris also are summarized. . Vol. 28; no. 2; pp. 211-231

Knotkova, D; Vlckova, J; Kreislova, K.
> Regional and microclimatic pollution effects on atmospheric corrosion in Prague and Europe. Materials Performance; Jun 30 1995. Description: Atmospheric corrosion test results for structural metals in open air and under shelter from multilateral European programs (1968 to 1992) are presented and the effects of SO_2 pollution analyzed. Regional and microclimatic pollution effects are documented by test results from Czech permanent test sites, test sites throughout Prague, and test sites on the St. Vitus Cathedral. The effects of variations in pollution activity, time of wetness, and type of metal are considered. Vol. 34; no. 6; pp. 41-47

Kong, F.X; Chen, Y.
> Effect of aluminum and zinc on enzyme activities in the green Alga Selenastrum capricorutum. Bulletin of Environmental Contamination and Toxicology; Nov 30 1995. Description: Acid rain produced by atmospheric pollution may decrease the pH value of water and increase the availability and potential toxicity of metals in water

which have detrimental effects on aquatic organism, including algae, the important component of the primary production, and, thus, the entire aquatic food chain. Recent reviews of the effects of acid rain on freshwater ecosystems have emphasized research interest in soluble trivalent aluminum, although Al is rated low among trace metals in biological importance. On the other hand, zinc is an important trace element for the growth of phytoplankton and the cofactor of some enzymes. Vol. 55; no. 5; pp. 759-765

Kortelainen, P.; David, M.B.; Roila, T.; Maekinen, I.
Acid-base characteristics of organic carbon in the HUMEX Lake Skjervatjern. Environment International; Jan 31 1992. Description: The Humic Lake Acidification Experiment (HUMEX) was launched in 1988 to study the role of humic substances in the acidification of surface waters and the impacts of acidic deposition on the chemical and biological properties of humic substances. This subproject was designed to determine the contribution of organic acids to the acidity of Lake Skjervatjern (the HUMEX Lake) and the impacts of the acidification on the characteristics of organic carbon. In order to get an empirical measure for organic acidity, dissolved organic carbon (DOC) was fractionated, isolated, and base-titrated from each half of Lake Skjervatjern. Vol. 186; pp. 621-629

Kouterick, K.B.; Skelly, J.M.; Pennypacker, S.P.; Cox, R.M.
Birch foliar responses to simulated acidic fog and *Septoria betulae* inoculations. Canadian Journal of Forest Research; Mar 01 2001. Description: The effects of simulated acidic fog and inoculation with *Septoria betulae* Pass. on foliar symptom development and foliar senescence of *Betula papyrifera* Marsh. and *Betula cordifolia* Regel seedlings were investigated in 1997 and 1998 under greenhouse conditions. An interactive role may exist between acidic fog events and *S. betulae* in causing birch foliar browning, a disease reported over the past decade to occur on mature trees growing adjacent to the Bay of Fundy, Canada. Vol. 31; no. 3; pp. 392-400

Kovacik, J.M.; Stoll, H.G.
The economics of repowering steam turbines. Proceedings of the American Power Conference; Jan 31 1990 Description: Repowering is defined as displacing steam presently generated in an existing fossil fuel fired boiler with a gas turbine-heat recovery steam generator (HRSG) system. The steam generated in the HRSG is expanded in the existing steam turbine generator. Repowering advantages include a significant increase in power output at an improved heat rate relative to the base value for the existing steam turbine cycle being repowered. Vol. 52; pp. 79-88

Kowalok, M.E.
Common threads: Research lessons from acid rain, ozone depletion, and global warming. Environment; Jul 31 1993. Description: Research on environmental hazards is often haphazard: Studies from different disciplines suddenly fit together in an unexpected way; scientists geographically far apart come to similar conclusions; and what initially appears to be a minor effect turns out to be critical. This brief account of the research leading to the discovery of three major environmental threats demonstrates that successful environmental research, despite its unpredictability, has several important characteristics. Vol. 35; no. 6; pp. 12-20

Kress, M.W.; Baker, R.; Ursic, S.J.
Chemistry response of two forested watersheds to acid atmospheric deposition Water Resources Bulletin; Oct 31 1990 Description: The deposition and chemistry of precipitation were estimated for one year in two forest ecosystems in the South-Central United States. Precipitation, throughfall, litter leachate, and soil leachate were analyzed for a small catchment of pine- hardwoods in southeastern Oklahoma and for a catchment of loblolly pines (Pinus taeda L.) in northern Mississippi. In the pine-

hardwood forest, 98 percent of the acid deposition was neutralized, 50 percent in the forest canopy, and 48 percent in the forest floor. In the pine forest, 75 percent of the acid deposition was neutralized, all in the forest floor. Vol. 265; pp. 747-756

Kreutzweiser, David. P.; Gunn, John. M.; Thompson, Dean. G.; Pollard, Heather. G.; Faber, Marvin. J.
Zooplankton community responses to a novel forest insecticide, tebufenozide (RH-5992), in littoral lake enclosures
Canadian Journal of Fisheries and Aquatic Sciences; Mar 01 1998
Description: The effects of tebufenozide (RH-5992), a potential forest insecticide, on zooplankton communities were determined in 16 littoral enclosures in a small forest lake of northern Ontario. Community structure in enclosures treated with 9, 36, or 157 µg tebufenozide/L (0.2, 0.7, and 3 times the expected environmental concentration) was compared with natural zooplankton communities in control enclosures. No significant treatment effects on zooplankton communities were detected, even at 3 times the expected environmental concentration. Vol. 55; no. 3; pp. 639-648

Kruger, J; Dean, M
Looking back on SO_2 trading: What's good for the environment is good for the market
Fortnightly; Aug 31 1997. Description: This article, written by Environmental Protection Agency (EPA) employees, documents patterns and trends in the trading of SO_2 allowances. EPA has developed a framework to investigate the SO_2 allowance market created under the Acid Rain Program. The system provides insight into the level and type of allowance trading activity. Reported transfers are placed into eight categories, which are then divided into three groups: arm's length transfers; intra-utility transfers, and reallocations. Analysis of over 2400 transfers moving approximately 38 million allowances between 1994 and 1997 are presented. Associated environmental benefits are also analyzed.
Vol. 135; no. 15; pp. 30-36

Krupa, Sagar.V.; Kickert, Ronald.N.
Considerations for establishing relationships between ambient ozone (O_3) and adverse crop response
Environmental Reviews; Jan 01 1997
Description: Exposures to the all pervasive ambient ozone (O_3) can and has resulted in visible foliar injury and (or) reduction of crop growth and yield. However, most of our knowledge regarding the latter effect is derived from above ambient, artificial O_3 fumigations in field exposure chambers. In the most recent years, such methodologies have been the subject of much criticism. Vol. 5; no. 1; pp. 55-77

Kullberg, A.; Petersen, R.C. Jr.; Hargeby, A.; Svensson, M.
Transport of octanol soluble carbon and dissolved organic carbon through the soil/water interface of the HUMEX lake
Environment International; Jan 31 1992
Description: The transport of dissolved organic carbon (DOC) and octanol soluble carbon (OSC) through the soil/water interface was studied at four hydraulic vents and four nonvents of the HUMEX experimental lake. These studies were performed during a summer 1990 (pre-treatment condition) and 1991 (post-treatment condition). The pH of the water coming from the hydraulic vents was on average 0.2-0.4 units lower than the pH of the lake water on the acid treated side of the lake. This difference was less on the control side. Vol. 186; pp. 631-636

Kulmala, M.; Laaksonen, A.
Binary nucleation of water--sulfuric acid system: Comparison of classical theories with different H_2SO_4 saturation vapor pressures. Journal of Chemical Physics; Jul 01 1990. Description: The classical hydrates interaction model and the classical condensation model have been compared in the temperature range 153.15--363.15 K. Various expressions for the kinetic part of nucleation rate have been examined. In the present study the saturation vapor pressure values of sulfuric acid given by Ayers *et al.* are extrapolated to lower (stratospheric) temperatures taking into account the temperature

dependence of the enthalpy of vaporization. In our model calculations we have compared the effect of three different expressions for saturation vapor pressures. The nucleation rate will differ by several orders of magnitude (depending on temperature, water and acid activity) when using different theories and/or saturation vapor pressures. Vol. 931; pp. 696-701

Kumar, Ashij J.; Gough, William A.; Karagatzides, Jim D.; Bolton, Kim A.; Tsuji, Leonard J. S.
Testing the Validity of a Critical Sulfur and Nitrogen Load Model in Southern Ontario, Canada, usingSoil Chemistry Data from MARYP. Environmental Monitoring and Assessment; Jul 01 2001. Description: The validity of a steady-state massbalance model (Arp et al., 1996; referred to asARP) was tested using physicochemical soil data fromthe Monitoring Acid Rain Youth Program (MARYP). FourARP sites were matched with ten MARYP sites accordingto proximity, bedrock type and subsoil pH to test thevalidity of the ARP model for critical loadexceedances. Vol. 69; no. 3; pp. 221-230

Kumar, K.S.; Feldman, P.L.; Jacobus, P.L.
Pulse energization: A precipitator performance upgrade technology following low sulfur coal switching. Proceedings of the American Power Conference; Jan 31 1992. Description: Madison Gas and Electric operates two 50 MWe pulverized coal fired boilers at its Blount station. This paper reports that these two units have been designed to operate with gas or coalfiring in combination with refuse derived fuel. Both these units are fitted with electrostatic precipitators for particulate control. Historically, these units have utilized Midwestern and Appalachian coals varying in sulfur contents between 2 and 5 %, with the SO_2 emission level in the 3.5 pounds per million Btu range. Wisconsin's acid rain control law goes into effect in 1993 requiring utilities to control sulfur dioxide emissions below 1.2 pounds per million Btu. Vol. 542; pp. 1181-1186

Kumar, N.; Kulshrestha, U.C.; Saxena, A.; Kumari, K.M.; Srivastava, S.S.
Formate and acetate in monsoon rainwater of Agra, India. Journal of Geophysical Research; Mar 20 1993. Description: Formate and acetate concentrations were estimated using ion chromatography in 19 precipitation samples collected on an eventwise basis during the monsoon season (July through September), 1991, at Dayalbagh, Agra. Volume-weighted average (VWA) concentrations for formate and acetate were 5.8 and 6.55 $[\mu]molL^{-1}$, respectively. The VWA hydrogen ion concentration was 0.084 $[\mu]eq\ L_{-1}$ (pH 7.07) and the correlation coefficient between the two ions was 0.85. The average formate to acetate ratio was low (0.88), possibly due to an increase in acetate contribution from direct emissions associated with heavy vehicular traffic load and/or indirect acetate formation by alkaline hydrolysis of PAN Vol. 983; pp. 5135-5137

Kumar, S.; C.V.S., Kameswara Rao,
Strength loss in concrete due to varying sulfate exposures. Cement and Concrete Research; Jan 31 1995. Description: The paper presents the results of an experimental study in which the loss in strength of concrete exposed to time varying sulfate environments is determined. A measure of cumulative damage to the material under time dependent exposure conditions is introduced. The test data indicate that the cumulative damage, when concrete is exposed to a succession of sulfate solutions of varying concentrations over successive time intervals, is dependent on the order of application of the exposure. Vol. 251; pp. 57-62

Kutka, F.J.
Low pH effects on swimming activity of Ambystoma salamander larvae. Environmental Toxicology and Chemistry; Nov 30 1994. Description: Swimming activity of larval Ambystoma laterale increased linearly with pH between pH 4.0 and 6.5; near inactivity occurred at pH 4.0. Ambystoma maculatum exhibited a weaker

relationship between activity and pH, and overall was less active than A. laterale. However, survival of larval A. maculatum declined linearly with pH in the presence of adult diving beetles (Dytiscus verticalis), with significantly lower survival below pH 4.8.
Vol. 1311; pp. 1821-1824

Kwong, K.V.; Meissner, R.E. III; Ahmad, S.; Wendt, C.J.
Application of amines for treating flue gas from coal-fired power plants. Environmental Progress; Aug 31 1991. Description: Recent advances have been made in hydrotreating of power plant boiler flue gases to reduce SO_x and NO_x, the major components of acid rain, to hydrogen sulfide and nitrogen, and to convert excess oxygen to water. The sulfur pollutants in the form of H_2S can be recovered and converted to elemental sulfur by a selective amine unit and a Recycle Seletox sulfur recovery unit.
Vol. 103; pp. 211-215

Lachance, Stephanie.; Bérubé, Pierre.; Lemieux, Michel.
In situ survival and growth of three brook trout (*Salvelinus fontinalis*) strains subjected to acid conditions of anthropogenic origin at the egg and fingerling stages. Canadian Journal of Fisheries and Aquatic Sciences; Aug 01 2000. Description: Tolerance to naturally acidic conditions of a Côte-Nord brook trout (*Salvelinus fontinalis*) strain (Arseneault strain, presumed acid tolerant) does not appear completely genetically mediated, since this tolerance was not evident when the fish were subject to acid conditions of anthropogenic origin. Three wild brook trout strains, at the egg and fingerling stages, were exposed in the field to waters acidified by atmospheric deposition of anthropogenic origin as well as to natural waters typical of the region. Although egg mortality was significantly higher in acid (61.0-85.6%) than in reference (6.3-20.8%) conditions, no differences between strains were noted.
Vol. 57; no. 8; pp. 1562-1573

Lambert, D.; McGowan, T.F.
NO_x control techniques for the CPI Chemical Engineering (New York); Jun 30 1996. Description: After years of air pollution control innovation, the control of emissions of nitrogen oxide compounds stands out as an area where much work remains to be performed in the chemical process industries (CPI). Federal regulations, ozone non-attainment areas, acid rain provisions of the US Clean Air Act, and corporate goals for emission reductions are all motivators. Primary CPI sources are high-temperature combustion systems, including fired heaters, boilers and Kilns. Nitrogen-based processes such as nitric acid manufacture also contribute. The paper discusses the regulations which define the problem and some solutions. These include fuel switching, low-excess air burners, fluegas recirculation, staged combustion, out of service burners, and wet scrubbing of flue gas. The paper briefly discusses costs of these options.
Vol. 103; no. 6; pp. 98-101

Latona, M.C; Neufeld, R.D; Hu, W; Kelly, C; Vallejo, L.E.
Response of MICROTOX organisms to leachates of autoclaved cellular concrete Journal of Energy Engineering; Aug 31 1997. Description: The MICROTOX bioassay, a toxicity test involving bioluminescent microorganisms, was conducted on aqueous leachates derived from a construction material made using coal fly ash as the key siliceous ingredient. The material is known as autoclaved cellular concrete (ACC). The test indicated an absence of toxic effects attributable to soluble species, which included the priority heavy metals in the filtered leachates. Toxic or inhibitive effects on the test bacteria were observed for the toxicity characteristic leaching procedure (TCLP) leachates, but this was probably due to acetic acid in the extractant rather than the solubilized metals.
Vol. 123; no. 2; pp. 35-43

Laurence, J.A.; Kohut, R.J.; Amundson, R.G.
Use of TREGRO to simulate the effects of ozone on the growth of red spruce

seedlings. Forest Science; Aug 31 1993
Description: TREGRO, a model developed to simulate the growth of sapling red spruce (Picea rubens Sarg.), was parameterized to grow 2- to 3-yr-old seedlings. Results of the simulation compared favorably to actual growth of seedlings used in a field study of the effects of ozone and acidic precipitation on tree physiology and development. Furthermore, a 10-yr simulation produced a modeled tree that corresponded to saplings used in another field experiment.
Vol. 393; pp. 453-464

Lave, L.; Gruenspecht, H.
Increasing the efficiency and effectiveness of environmental decisions: Benefit- cost analysis and effluent fees: A critical review Journal of the Air and Waste Management Association; Jun 30 1991
Description: The objective of this critical review is to evaluate and summarize the literature on economic tools used to improve environmental decision making. Environmental decisions are complicated by pervasive uncertainty, and the lack of consensus on goals and on tradeoffs, such as air versus water pollution and versus loss of coal mining jobs. Five decision frameworks are used by regulatory agencies to simplify decision making.
Vol. 416; pp. 680-693

Lawson, D.R.
The Southern California air quality study Journal of the Air and Waste Management Association; Feb 28 1990. Description: During the 82nd Annual Air Waste Management Association meeting in Anaheim, California, researchers presented preliminary results from the 1987 Southern California Air Quality Study (SCAQS), which was conducted in the South Coast Air Basin. This paper describes the background and study design for SCAQS, presents summaries of the papers presented in the SCAQS Symposium at the A WMA meeting, and describes the data analysis and air quality modeling efforts currently in progress as part of the SCAQS program.
Vol. 402; pp. 156-16

Leaf, D.A.
Acid rain and the Clean Air Act Chemical Engineering Progress; May 31 1990. Description: This article focuses on acid rain and discusses some background information, the existing Clean Air Act, and the key provisions of the president's proposal to amend the Clean Air Act. The discussion of the proposal includes the size and timing of reductions, the affected sources, and the various approaches and technologies likely to be used to achieve the emission reductions.
Vol. 865; pp. 25-29

Lee, B.
Highlights of the Clean Air Act Amendments of 1990. Journal of the Air and Waste Management Association; Jan 31 1991. Description: The amendments to the Clean Air Act provide for operating permits for stationary sources of pollution, marking a change that is comparable to federal water pollution regulations. The bill addresses acid rain emissions and will phase out production of chemicals contributing to depletion of the stratospheric ozone layer. A major new concept incorporated into the emission limits established by law is a system of tradeable emissions credits. If a facility reduces emissions below the standard or ahead of the timetable set by law, emissions credits are earned that can be applied to future emissions or sold to another facility. The bill also provides strict deadlines for the EPA to meet in promulgating the regulations. These deadlines will seriously strain the personnel and financial resources of the agency. Perhaps the most significant of the miscellaneous items in the law is a provision for extended unemployment benefits under the Job Training Partnership Act for workers who lose their jobs because of the law's provisions. However, this is contingent on the displaced workers' seeking job retraining.
Vol. 411; pp. 16-19

Lee, W.S.; Chevone, B.I.; Seiler, J.R.
Growth response and drought susceptibility of red spruce seedlings exposed to

simulated acidic rain and ozone Forest Science; Jun 30 1990. Description: One-year-old seedlings of red spruce were exposed to O_3 ({le} 0.025 or 0.10 {mu}l l^{-1}, 4 h d^{-1}, 3 d wk^{-1}) in combination with simulated rain (pH 5.6 or 3.0, 1 h d^{-1}, 2 d wk^{-1}, 0.75 cm h^{-1}) for 10 weeks. After pollutant treatments, seedlings were subjected to two successive drought cycles. Whole-plant fresh weight increment (FWT) and dry weight were reduced after O_3 exposure, whereas FWT and shoot height growth were increased after simulated rain exposure at pH 3.0 compared to pH 5.6. No interaction between O_3 and rain treatments was observed for any growth variable measured. Foliar concentrations of K and S were greater in seedlings exposed to simulated rain at pH 3.0 compared with those at pH 5.6. Root hydraulic conductivity was highest in seedlings exposed to 0.10 {mu}l l^{-1} O_3 + pH 3.0 rain solution compared with all other treatments after the first drought cycle. Vol. 362; pp. 265-275

LeMay, B.; Willyard, B.; Polasek, S.; Clarkson, C.W.
Design, maintenance extend FGD system slurry valve life. Power Engineering (Barrington); Aug 31 1995. Description: This article describes how power plants in Florida, Oklahoma and Texas adopted improved maintenance techniques and sought better design criteria to gain greater slurry valve reliability. Slurry valves, a vital part of a flue gas desulfurization (FGD) system, are critical to a power plant's ability to meet or exceed acid rain emission requirements. The performance and reliability of these valves can significantly affect unit operation and load capacity. Vol. 99; no. 8; pp. 45-48

Lenglet, M; Lopitaux, J.; Leygraf, C; Odnevall, I.; Carballeira, M; Noualhaguet, J.C.; Guinement, J; Gautier, J.; Boissel, J
Analysis of corrosion products formed on copper in $Cl_2/H_2S/NO_2$ exposure Journal of the Electrochemical Society; Nov 30 1995. Description: A joint effort has been undertaken, involving four laboratories and nine different analytical techniques, aiming at characterizing corrosion products on copper formed in mixed flowing gas test containing Cl_2, H_2S, and NO_2. The results and considerations clearly show that the identification of different crystalline or noncrystalline phases in the corrosion products only can be performed by a combination of independently working analytical techniques. Cuprite, Cu_2O, is the dominating phase during initial exposure, whereas cuprite together with atacamite, $Cu_2Cl(OH)_3$, are the most abundant phases after 10 days of exposure.
Vol. 142; no. 11; pp. 3690-3696

Levin, M.H.
Environmental trends. Journal of the Air and Waste Management Association; Feb 28 1990. Description: This article discusses acid rain legislation designed to achieve a 10-million ton annual SO_2 reduction from 1980 levels. H.R. 3030, Title V, though achieving the SO_2 reduction also capped each units' total emissions, generally based on actual fuel use in 1985-87 and granted it tradeable credits or allowances only for tons of emissions equal to that cap. New utility units (including cogenerators or independent power producers commencing operation after the Act's passage) received no allowances but were required to secure offsets from others by 2001. Other procedural constraints, and complex permit requirements have drawn charges that the bill would restrict normal utility load shifts, grandfather most allowances to high-emitting midwest coal plants, penalize clean states, stifle growth and drive future generation from coal to gas. Both these changes and their rebuttal revolve around three issues: cost, cap and complexity. All are related to how much trading will occur under the cap.
Vol. 402; pp. 144, 263

Levin, M.H.
Environmental trends: The Tox-man cometh. Journal of the Air and Waste Management Association; Aug 31 1990 Description: The Air Toxics Title of the pending Clean Air Act could pose

problems for businesses of all sizes. While public attention largely focused on Congressional battles over acid rain and urban smog, tough clauses to eliminate hazardous emissions from production or commercial use of such common substances as paper, plastic, antifreeze, methanol and glass wool wailed serenely by. As passed in similar versions by the Senate and House, scheduled for summer action by a Conference Committee, and likely to be signed into law next October, these clauses will sweep thousands of businesses. They could subject dry cleaners as well as factories to neighborhood hearings on their continued existence, and to criminal penalties for failure to keep detailed records or prevent certain emissions that have not yet been regulated. They could also cost from $7 billion to more than $23 billion per year for compliance and endanger nearly a million net private-sector jobs upon full implementation, making air toxics rules one of the most expensive environmental programs ever enacted.
Vol. 408; pp. 1084, 1148

Levine, J.S.
The consequences of global biomass burning. Earth in Space; Jan 31 1991. Description: Global biomass burning encompasses forest burning for land clearing, the annual burning of grasslands, the annual burning of agricultural stubble and waste after harvests, and the burning of wood as fuel. These activities generate CO_2, CH_4 and other hydrocarbons, CO, H_2, NO, NH_3, and CH_3Cl; of these, CO, CH_4 and the hydrocarbons, and NO, are involved in the photochemical production of tropospheric O_3, while NO is transformed to NO_2 and then to nitric acid, which falls as acid rain. Biomass burning is also a major source of atmospheric particulates and aerosols which affect the transmission of incoming solar radiation and outgoing IR radiation through the atmosphere, with significant climatic effects. Vol. 3; pp. 5-7

Lin, Teng-Chiu.; Hamburg, Steven.P.; Hsia, Yue-Joe.; King, Hen-Biau.; Wang, Lih-Jih.; Lin, Kuo-Chuan.
Base cation leaching from the canopy of a subtropical rainforest in northeastern Taiwan. Canadian Journal of Forest Research; Jul 01 2001. Description: We examined base cation leaching from the canopy of a subtropical rainforest in northeastern Taiwan. The forest is characterized by extremely low levels of base cations in both canopy vegetation and in the soils. The rates of canopy leaching of K^+, Ca^{2+}, and Mg^{2+} were very high, representing up to 30, 35, and 190%, respectively, of the amount stored in leaves. Vol. 31; no. 7; pp. 1156-1163

Lindberg, S.E.; Bredemeier, M.; Schaefer, D.A.; Qi, L.
Atmospheric concentrations and deposition of nitrogen and major ions in conifer forests in the United States and Federal Republic of Germany. Atmospheric Environment; Jan 31 1990. Description: Emission densities of air pollutants are higher in Europe than in the US as a whole, suggesting similar differences in atmospheric deposition. The authors determined air concentrations and deposition during the warm season at conifer forests in Tennessee and northern Germany. The authors results confirmed major differences in both chemistry and fluxes. Vol. 248; pp. 2207-2220

Linden, H.R.
Prophecies of doom. Fortnightly; Sep 15 1993. Description: Biased interpretations of environmental data are choking the life out of the global energy system. Prevalent biases in the interpretation of scientific, technical, resource, and economic data influence energy policy and define the new energy/environment/economy (E^3) paradigm. These biases impede the orderly evolution of the global energy system to sustainability because they make rising energy consumption appear far less socially, economically, & environmentally beneficial than it is in fact - both on the basis of historical record and of forecasts that assign proper weight to ongoing technology advances.
Vol. 13117; pp. 13-15

Ling, K.A.; Ashmore, M.R.; Macrory, R.B.
The use of word-based models to describe the development of UK acid rain policy in the 1980s. Environmental Science and Policy; Oct 01 2000. Description: Understanding how environmental policy decisions were reached in the past might help predict policy development in the future. This paper evaluates how well two existing frameworks for decision analysis fit acid rain policy development of the UK Central Electricity Generating Board (CEGB) in the 1980s. Decision tree analysis assumes a rational approach to decision-making and overlooks the dynamic nature of the decision making process. Vol. 3; no. 5; pp. 249-262

Lobsenz, G
EEI contests N.Y. concerns with allowance trading scheme. Energy Daily; Jul 07 1993. Description: A utility industry analyst has challenged charges that sulfur dioxide emissions allowance trading may be environmentally damaging, citing studies showing no harmful increases in acid deposition due to simulated interstate allowance transactions. Contrary to allegations by New York state officials, John Kinsman of the Edison Electric Institute said existing studies indicate the Adirondacks and other sensitive Northeast ecosystems actually may benefit from SO_2 emissions allowance trading. In particular, studies done by the National Acid Precipitation Assessment Program suggest allowance trading could marginally reduce acid deposition in the Adirondacks below levels already expected under the federal SO_2 emissions reduction program for utilities, said Kinsman, an environmental scientists at EEI. Vol. 21; no. 128; pp. 1, 3

Lobsenz, G
Illinois Power, PacifiCorp protest EPA acid rain plan. Energy Daily; Sep 07 1993. Description: In language clearly threatening legal action, PacifiCorp and Illinois Power have lodged a strong protest with the EPA over its decision to defer action on utility Phase I acid rain substitution plans. They specifically complained about EPA's deferral of a plan under which PacifiCorp's Gadsby Unit 3, Wyodak Unit BW91 and Jim Bridger Units BW71, 72, and 73 would serve as substitution for Illinois Power's Baldwin Unit 2 for 1996-99. This was one of more than 75 plans deferred by the EPA. Vol. 21; no. 171; pp. 1, 4

Lobsenz, G
State inaction could boost clean air costs - EPA. Energy Daily; Jul 12 1994. Description: There is an "urgent" need for state utility commissions to issue ratemaking guidance on acid rain allowances because the "dearth" of regulatory action is stifling emissions trading, the Environmental Protection Agency says. Vol. 22; no. 131; pp. 1, 4

Lobsenz, G.
Clean air compliance option of choice is low-sulfur coal. Energy Daily; Apr 18 1994. Description: Lower-than-expected prices for low-sulfur coal made fuel-switching the compliance strategy of choice for utilities in Phase I of the acid rain program, with 62 percent of affected power plants selecting that option, the Energy Information Administration reported. In a report detailing utilities' Phase I compliance strategies, 162 of the 261 units facing 1995 emissions reduction deadlines were meeting their obligations by buying low-sulfur coal. Another 39 units will come into compliance by buying additional sulfur dioxide emissions allowances, mostly from other utilities. Vol. 2272; pp. 3

Lobsenz, G.
Court date for EPA acid rain rule. Energy Daily; Mar 04 1994. Description: In an acid rain rulemaking that appears headed straight for the courtroom, the Environmental Protection Agency this week announced new limits on emissions of nitrogen oxides from coal-fired power plants. The regulations, announced March 1, are expected to achieve a 1.8 million ton per year reduction in power plant NO_x emissions, which are considered a major contributor to acid rain. The agency issued companion regulations last year to cut

power plant discharges of sulfur dioxide, the other major acid rain pollutant. Vol. 2242; pp. 3

Lobsenz, G.
EDF: New York oversight plan a disaster for SO$_2$ trades. Energy Daily; Mar 16 1993. Description: New York state officials have asked a federal court to review the Environmental Protection Agency's sulfur dioxide emission allowance trading program, charging that it could allow Midwest utilities to discharge excessive acid rain pollution. In filing a lawsuit against EPA in the US Court of Appeals for the District of Columbia Friday, the state claimed the agency failed to establish a proper "oversight mechanism" for allowance trades. Vol. 2150; pp. 1-2

Lobsenz, G.
EPA rates second allowance auction a success. Energy Daily; Mar 30 1994. Description: The Environmental Protection Agency's second annual auction of sulfur dioxide emission allowances drew more bids than last year's auction, but average winning prices remained flat at around $150 to $160. The auction also clearly indicated growing market interest in Phase II allowances usable for acid rain compliance after 2000. Of 284 total bids, 181 were for six-year advance or seven-year advance allowances while 103 were "spot" allowances usable for Phase I compliance starting in 1995. That represented a reverse from EPA's first auction last year when there were 106 bids for spot allowances and 65 for advance allowances. Vol. 2260; pp. 1, 4

Lobsenz, G.
Inexpensive allowances may be costly Energy Daily; Jun 21 1993
Description: Utilities that install scrubbers to meet acid rain reduction requirements may not recover their investments due to rapidly declining prices for sulfur dioxide allowances after 2000, a new study warns. SO$_2$ allowance prices probably will peak around the year 2000 at about $500 per ton; they then will decline to $100 by 2015 before falling close to zero by 2020, according to the study by five researchers at the New York State Energy Office. Vol. 21117; pp. 1-2

Lobsenz, G.
Opting for tried and true, utilities are overpaying for SO$_2$ control-NRRI Energy Daily; Mar 23 1994
Description: Utilities are incurring much higher acid rain control costs than necessary due to their failure to participate in the so-far bargain-basement sulfur dioxide emissions allowance market, according to a new National Regulatory Research Institute study. The NRRI study said utilities included in Phase 1 of the federal acid rain reduction program overwhelmingly have opted for fuel switching or scrubbers to meet their pollution control requirements, even though those cleanup options cost up to six times more per ton of SO$_2$ than the market cost of allowances. Vol. 2255; pp. 1, 3

Lobsenz, G.
Second EPA allowance auction set for today. Energy Daily; Mar 28 1994. Description: Utilities will generate substantial surplus sulfur dioxide emissions allowances in Phase 1 of the federal acid rain program, but they generally will bank those excess allowances for their own internal use in Phase 2, according to a recently released study. The Electric Power Research Institute study, suggests that private trading activity may be relatively low in the Environmental Protection Agency second annual allowance auction. Allowance market experts believe that, as in the first auction in March 1993, few of the private allowances on the trading block this year actually will be sold. Vol. 2258; pp. 1, 2

Lock, R.
Utility regulation and the Clean Air Act Amendments of 1990
Proceedings of the American Power Conference; Jan 31 1991
Description: The focus of this paper is on acid rain and the interaction with state and federal economic regulation, i.e. public

utility regulation as carried out by state PUCs and by FERC. While some of those other sections, especially where they interact with Title IV, may affect state and federal regulation in the long run, most of the regulatory action under those sections will be at the EPA level, usually utilizing traditional command and control standards and enforcement mechanisms.
Vol. 53; pp. 321-325

Loeb, A.P.
Addressing the public's goals for environmental regulation when communicating acid rain allowance trades Electricity Journal; May 31 1995 Description: The criticisms that followed the first allowance trades remain a mystery - why, if trading produces efficiencies, has the public been critical? Perhaps there is a better reason to support trading.
Vol. 8; no. 4; pp. 55-63

Loerting, Thomas; Liedl, Klaus R.
Toward elimination of discrepancies between theory and experiment: The rate constant of the atmospheric conversion of SO3 to H2SO4. U.S. Proceedings of The National Academy of Sciences; Aug 1 2000. Description: The hydration rate constant of sulfur trioxide to sulfuric acid is shown to depend sensitively on water vapor pressure. In the 1:1 SO3-H2O complex, the rate is predicted to be slower by about 25 orders of magnitude compared with laboratory results.
Vol. 97; no. 16; pp. 8874-8878

Long, J.
Farm, air, and budget bills await action Chemical and Engineering News; Oct 29 1990. Description: This article describes farm, air, and budget legislation awaiting action by the 101st U.S. Congress. Included are bills on funding of the Superconductor Supercollider, the development of a national supercomputer network, and amending the Clean Air Act to reduce emissions of acid rain precursors and toxic air pollutants.
Vol. 6844; pp. 18-19

Long, R.Q.; Yang, R.T.
Superior Fe-ZSM-5 catalyst for selective catalytic reduction of nitric oxide by ammonia
Journal of the American Chemical Society; Jun 16 1999
Description: Nitrogen oxides in the exhaust gases from combustion of fossil fuels remain a major source for air pollution and acid rain. The current technology for reducing NO_x ($NO + NO_2$) emissions from power plants is selective catalytic reduction (SCR) with ammonia in the presence of oxygen. For the SCR reaction, $V_2O_5 + WO_3$ (or MoO_3) supported on TiO_2 are the commercial catalysts.
Vol. 121; no. 23; pp. pp.5595-5596

Loucks, O.L.
Science or politics NAPAP and Reagan Forum for Applied Research and Public Policy; Jan 31 1993. Description: The National Acid Precipitation Assessment Program (NAPAP) was used during the Reagan administration as much to serve political as scientific ends, charges Orie L. Loucks of Miami University in Oxford, Ohio. "An apparent goal was to have the scientific assessments support a pre-determined decision: to postpone action on acid-gas emissions controls." Loucks compares NAPAP efforts to compute the costs and expected benefits of acid-rain cleanup with the "slanted intelligence" provided by the Central Intelligence Agency under William Casey - a key figure in the "Irangate" controversy. Vol. 82; pp. 66-73

Lovett, G.M.
Atmospheric deposition of nutrients and pollutants in North America: An ecological perspective. Ecological Applications; Nov 30 1994. Description: Research on air pollution and acidic deposition in the last 15 yr has greatly increased knowledge of the rates and the processes of atmospheric deposition. Invigoration of the field has been a direct result of interchange and cooperation among ecosystem ecologists, micrometeorologists, and plant physiological ecologists who approach atmospheric deposition from different

perspectives. There is a widespread realization among ecologists of the importance of dry and cloud deposition and the introduction of new methods to estimate these fluxes. Vol. 44; pp. 629-650

Lucas, A
Pollution trading rights could spread Chemical Week; Apr 06 1994 Description: Bearing out the concept that market-based solutions are efficient in reducing pollution, The Chicago Board of Trade (CBOT) and EPA say the second annual acid rain emission auction was successful in making strides toward reducing acid rain. Futhermore, that success may lead to expanding emissions trading to other industries and other emissions. Vol. 154; no. 13; pp. 16

Luley, C.J
The greening of urban air Forum for Applied Research and Public Policy; Jan 31 1998. Description: Trees growing in urban areas can make significant contributions to urban air quality in addition to beautifying downtown zones. The reason: trees and other plants absorb such pollutants as ozone and nitrogen oxides. As a consequence, incentives to plant and manage urban trees could be integrated into federal and state air-quality programs as part of a long-range strategy for combating urban air pollution.
Vol. 13; no. 2; pp. 33-35

Lux, Heidi.B.; Cumming, Jonathan.R.
Mycorrhizae confer aluminum resistance to tulip-poplar seedlings. Canadian Journal of Forest Research; Apr 01 2001 Description: Aluminum (Al) toxicity may limit the growth and nutrient acquisition of sensitive tree species in regions receiving acidic deposition. Symbioses between tree roots and mycorrhizal fungi may offset the negative impacts of Al in the root zone. *Liriodendron tulipifera* L. (tulip-poplar) is an important tree species in the Appalachian Mountains of the southeastern United States and may be at risk from the high levels of acidic deposition in that area. Mycorrhizal and non-mycorrhizal tulip-poplar seedlings were exposed to Al levels of 0, 50, 100, and 200 µM in sand culture for 6 weeks. Vol. 31; no. 4; pp. 694-702

Lydersen, E.; Henriksen, A.
Total organic carbon in streamwater from four long-term monitored catchments in Norway. Environment International; Jan 31 1994. Description: Flux, concentration, and net charge of total organic carbon (TOC) have been related to different physico-chemical parameters present in air/precipitation and streamwater at four long-term monitored catchments in Norway, during the period 1986-1992. The catchments vary a lot with respect to annual water input, acid rain, and the streamwater concentration of TOC. Thus, relationships between concentration/NC of TOC and chemical compounds in precipitation and streamwater were often catchment-specific. Vol. 206; pp. 713-729

Lydersen, E.; Salbu, B.; Poleo, A.B.S.; Muniz, I.P.
Formation and dissolution kinetics of $Al(OH)_3(s)$ in synthetic freshwater solutions. Water Resources Research; Mar 31 1991. Description: The precipitation of Al in aqueous solutions can be described as a two-step process. When acidic inorganic Al solutions (pH 4.5) were titrated with NaOH to pH levels between 5.5 and 6.0, an amorphous $Al(OH)_3$ (s) phase was formed instantaneously. During the first 5 min, the apparent half time for the reduction in dissolved Al species (t $\{1/2\}$) was 0.162 ± 0.07 hours (n = 4). The decrease of dissolved Al species continued during the following 24 hours, but at a far slower rate. The highest precipitation rates were found in the solution of highest pH, and at approximately identical pH, the highest rate was found in the solution of highest temperature. Vol. 273; pp. 351-357

Lynch, J.A.; Bowersox, V.C.; Grimm, J.W.
Acid rain reduced in Eastern United States Environmental Science and Technology; Mar 15 2000. Description: Concentrations of sulfate (SO_4^{2-}) and free hydrogen ions (H^+) in precipitation decreased from 10% to 25% over a large area of the Eastern US

from 1995 through 1997 as compared to the previous 12-year (1983--1994) reference period. These decreases were unprecedented in magnitude and spatial extent. In contrast, nitrate (NO_3^-) concentrations generally didn't change over this period. Vol. 34; no. 6; pp. 940-949

MacDonald, N.W.; Hart, J.B. Jr.
Relating sulfate adsorption to soil properties in Michigan forest soils. Soil Science Society of America Journal; Jan 31 1990. Description: Six Michigan forest soil series comprising several gradients in soil physical and chemical properties were studied to relate SO_4 adsorption to soil properties. The series studied were Grayling (mixed, frigid Typic udipsamments), Rubicon (sandy, mixed, frigid Entic Haplorthods), Kalkaska (sandy, mixed, frigid Typic Haplorthods), Montcalm (coarse-loamy, mixed Eutric Glossoboralfs), Spinks (sandy, mixed, mesic Psammentic Hapludalfs), and Oshtemo (coarse-loamy, mixed, mesic Typic Hapludalfs). Six randomly located map units of each series were sampled in the lower peninsula of Michigan. An index of SO_4 adsorption potential was determined by shaking samples in solutions containing 0.312 mmol SO_4-S L^{-1} and measuring the loss of SO_4 from solution. The highest SO_4-adsorption potentials were found in Bw, Bs, and Bhs horizons of the Grayling, Rubicon, Kalkaska, and Montcalm series (7.0 to 14.1 mg S kg^{-1}) and in E and Bt horizons of the Spinks and Oshtemo series (4.3 to 10.1 mg S kg^{-1}). Vol. 541; pp. 238-245

MacDonald, N.W.; Witter, J.A.; Burton, A.J.; Pregitzer, K.S.; Liechty, H.O.; Mroz, G.D.
Ion leaching in forest ecosystems along a Great Lakes air pollution gradient
Journal of Environmental Quality; Jan 31 1992. Description: A gradient of H^+, SO_4^{2-}, and NO_3^- deposition across the Great Lakes region raised concerns over impacts on soil solution chemistry and ion leaching in regional forest ecosystems. Ten study sites representing northern hardwood and oak ecosystems were established across the gradient of increasing deposition from Minnesota to Ohio. Lysimeters were installed at lower E and lower B horizon boundaries at each site and sampled over a 2-yr period. Vol. 214; pp. 614-623

Mackun, I.R.; Leopold, D.J.; Raynal, D.J.
Short-term responses of wetland vegetation after liming of an Adirondack watershed Ecological Applications; Aug 31 1994 Description: Watershed liming has been suggested as a long-term mitigation strategy for lake acidity, particularly in areas subject to high levels of acidic deposition. However, virtually no information has been available on the impacts of liming on wetland vegetation. In 1989, 1100 Mg of limestone (83.5% $CaCO_3$) were aerially applied to 48% (100 ha) of the Woods Lake watershed in the west-central Adirondack region of New York as part of the first comprehensive watershed liming study in North America. Vol. 43; pp. 535-543

MacLeod, C.; Taylor, E.
Effects of infiltration chemistry on the mobilization potential of mercury (Hg) in soils from the New Jersey Coastal Plain Geological Society of America, Abstracts with Programs; Jan 31 1992. Description: Mercury concentrations in ground water exceeding the USEPA maximum contaminant level of 2 [mu]g/l have been found in wells 60 to 100 feet deep in the Kirkwood- Cohansey Aquifer System of the New Jersey Coastal Plain. The aquifer is a sand and gravel aquifer consisting primarily of quartz with minor amounts of biotite, plagioclase feldspars and ilmenite. This is the shallowest aquifer system in the NJ Coastal Plain and is unconfined over much of central and southern NJ. Soils on these aquifer sediments are quartz-rich with poorly developed A_0 and B horizons. Vol. 247; pp. A79

MacQuarrie, D.M.; Mavinic, D.S.; Neden, D.G.
Greater Vancouver Water District drinking water corrosion inhibitor testing Canadian Journal of Civil Engineering; Feb 01 1997. Description: This study was initiated to evaluate, in Greater Vancouver Water District water, the effectiveness of

zinc orthophosphate, type N sodium silicate, and a commercial blend of the two as corrosion inhibitors within the limitations that a seven-loop pilot plant would allow. In all but the raw water control loop, pH and alkalinity were adjusted, and the water was disinfected with 2.5 mg/L of chloramine (NH_2Cl). Copper and cast iron corrosion rates were measured over the course of 12 months on pipe inserts removed at 3-month intervals and via weekly monitoring using an electrical resistance measuring device. Vol. 24; no. 1; pp. 34-52

Madigosky, S.R.; Alvarez-Hernandez, X; Glass, J.
Concentrations of aluminum in gut tissue of crayfish (Procambarus clarkii), purged in sodium chloride
Bulletin of Environmental Contamination and Toxicology; Oct 31 1992
Description: Recent concern over the release of Al in the environment has prompted researchers and health officials to assess its effects on biological systems. Aluminum, despite being the most abundant metal in earth's lithosphere, is normally complexed in soil and is therefore unavailable for biological assimilation. The recent advent of acid rain, however, has prompted Al release due to mobilization from surrounding sediments into the environment. Vol. 49; no. 4; pp. 626-632

Madsen, B.C.; Dreschel, T.W.
Acid rain monitoring in Florida from 1978 to the present and evaluation of trends in rainwater composition. National Meeting - American Chemical Society, Division of Environmental Chemistry; Jan 31 1996. Description: The occurrence of acid rain has prompted extensive research and monitoring activities which began in the U.S. during the late 1970's. In the mid 1970's the National Aeronautics and Space Administration (NASA) funded an extensive environmental monitoring program which included a substantial acid rain monitoring component. Results from that study and subsequent activities have been summarized in previous reports. Vol. 36; no. 2; pp. 52-55

Maibodi, M.
Implications of the Clean Air Act acid rain title on industrial boilers. Environmental Progress; Nov 30 1991. Description: This paper discusses the impacts of the 1990 Clean Air Act Amendments related to acid rain controls, as they apply to industrial boilers. Emphasis is placed on explaining the Title IV provisions of the Amendments that permit nonutility sources to participate in the SO_2 allowance system. The allowance system, as it pertains to industrial boiler operators, is described, and the opportunities for operators to trade and/or sell SO_2 emission credits is discussed. The paper also reviews flue gas desulfurization system technologies available for industrial boiler operators who may choose to participate in the system. Furnace sorbent injection, advanced silicate process, lime spray drying, dry sorbent injection, and limestone scrubbing are described, including statements of their SO_2 removing capability, commercial status, and costs. Capital costs, levelized costs and cost-effectiveness are presented for these technologies. Vol. 104; pp. 307-313

Malcolm, R.L.
^{13}C-NMR spectra and contact time experiment for Skjervatjern fulvic and humic acids. Environment International; Jan 31 1992. Description: The T_{CP} and $T_1[rho]$ time constants for Skjervatjern fulvic and humic acids were determined to be short with T_{CP} values ranging from 0.14 ms to 0.53 ms and $T_1[rho]$ values ranging from 3.3 ms to 5.9 ms. T_{CP} or $T_1[rho]$ time constants at a contact time of 1 ms are favorable for quantification of ^{13}C-NMR spectra. Because of the short T_{CP} values, correction factors for signal intensity for various regions of the ^{13}C-NMR spectra would be necessary at contact times greater than 1.1 ms or less than 0.9 ms. T_{CP} and $T_1[rho]$ values have a limited non-homogeneity within Skjervatjern fulvic and humic acids. A pulse delay or repeat time of 700 ms is more than adequate for quantification of these ^{13}C-NMR spectra. Vol. 186; pp. 609-620

Malcolm, R.L.; Hayes, T.
  Organic solute changes with acidification in Lake Skjervatjern as shown by ^{1}H-NMR spectroscopy. Environment International; Jan 31 1994. Description: ^{1}H-NMR spectroscopy has been found to be a useful tool to establish possible real differences and trends between all natural organic solute fractions (fulvic acids, humic acids, and XAD-4 acids) after acid-rain additions to the Lake Skjervatjern watershed. The proton NMR technique used in this study determined the spectral distribution of nonexchangeable protons among four peaks (aliphatic protons; aliphatic protons on carbon α or attached to electronegative groups; protons on carbons attached to O or N heteroatoms; and aromatic protons).
  Vol. 203; pp. 299-305

Malcolm, R.L.; MacCarthy, P.
  Quantitative evaluation of XAD-8 and XAD-4 resins used in tandem for removing organic solutes from water Environment International; Jan 31 1992 Description: The combined XAD-8 and XAD-4 resin procedure for the isolation of dissolved organic solutes from water was found to isolate 85% or more of the organic solutes from Lake Skjervatjern in Norway. Approximately 65% of the dissolved organic carbon (DOC) was first removed on XAD-8 resin, and then an additional 20% of the DOC was removed on XAD-4 resin. Approximately 15% of the DOC solutes (primarily hydrophilic neutrals) were not sorbed or concentrated by the procedure. Vol. 186; pp. 597-607

Mallory, M.L.; McNicol, D.K.; Cluis, D.A.; Laberge, C.
  Chemical trends and status of small lakes near Sudbury, Ontario, 1983-1995: evidence of continued chemical recovery Canadian Journal of Fisheries and Aquatic Sciences; Jan 01 1998. Description: We monitored 23 chemical parameters in 161 lakes northeast of Sudbury, Ontario, in most years between 1983 and 1995 to determine whether lake chemistries were responding to reduced local SO_2 emissions. Lakes were typically small (median 4.0 ha, 4.5 m deep), rapid flushing, and acid stressed (median pH 5.58, acid-neutralizing capacity (ANC) 7.1 μequiv.·L^{-1}). Forty percent of the lakes declined significantly in SO_4, base cations, and Al levels from 1983 to 1995, but only 12 and 16% increased in ANC and pH, respectively.
  Vol. 55; no. 1; pp. 63-75

Mänd, Raivo.; Tilgar, Vallo.; Leivits, Agu.
  Reproductive response of Great Tits, *Parus major*, in a naturally base-poor forest habitat to calcium supplementation Canadian Journal of Zoology; May 01 2000. Description: Recent studies have revealed that some forest passerines have difficulty obtaining sufficient calcium (Ca) for their eggshells in heavily acidified areas. However, the effect of Ca limitation on breeding success in non-acidified but naturally base-poor breeding habitats is not yet clear. The issue itself is important, insofar as the cost of egg formation in a certain habitat depends on the availability not only of energy and proteins but also of shell-formation material.
  Vol. 78; no. 5; pp. 689-695

Markewitz, D; Richter, D.D.; Allen, H.L.; Urrego, J.B.
  Three decades of observed soil acidification in the Calhoun Experimental Forest: Has acid rain made a difference? Soil Science Society of America Journal; Sep 30 1998. Description: Three decades of repeated soil sampling from eight permanent plots at the Calhoun Experimental Forest in South Carolina allowed the authors to estimate the rate of soil acidification, the chemical changes in the soil exchange complex, and the natural and anthropogenic sources of acidity contribution to these processes. During the first 34 yr of loblolly pine (Pinus taeda L.) forest growth, soil pH, decreased by 1 unit in the upper 0- to 15-cm of soils and by 0.4 and 0.3 units in the 15- to 35- and 35- to 60-cm layers, respectively. Throughout the 0- to 60-cm horizon, base cation depletion averaged 1.57 $kmol_c$ ha^{-1} yr^{-1} and effective and total acidity increased by 1.26 and 3.28 $kmol_c$ ha^{-1} yr^{-1}, respectively.
  Vol. 62; no. 5; pp. 1428-1439

Markewitz, D; Richter, D.D; Allen, H.L.
An H^+ ion budget approach for evaluating three decades of forest soil acidification
Bulletin of the Ecological Society of America; Jun 30 1995
Description: The acidity of a forest soil affects a wide range of ecological processes and is in turn affected by many of these processes. Hydrogen ion budgets may be used to integrate the array of processes that interact to produce and consume forest soil acidity. Three decades of forest soil observation, in the 0-60 cm layer, in a loblolly pine plantation, reports a mean annual acidification, in KCl-exchangeable acidity, of 1.33 kmolc/ha. During this period soil pH decreased by as much as one pH unit.
Vol. 76; no. 2; pp. 167

Markey, E.J.; Moorhead, C.J.
The Clean Air Act and bonus allowances
Public Utilities Fortnightly; May 15 1991
Description: This article discusses how utility companies can benefit in the form of bonus sulfur dioxide allowances from the Environmental Protection Agency by investing in renewable energy sources such as wind and promoting conservation. Topics discussed include the Clean Air Act Amendments, acid rain, energy conservation, renewable energy sources, and the procedure for gaining bonus allowances. Vol. 12710; pp. 30-34

Martin, C.W.; Driscoll, C.T.; Fahey, T.J.
Changes in streamwater chemistry after 20 years from forested watersheds in New Hampshire, U.S.A. Canadian Journal of Forest Research; Aug 01 2000.
Description: Long-term patterns of streamwater chemistry provide valuable evidence of the effects of environmental change on ecosystem biogeochemistry. Observations from old-growth forests may be particularly valuable, because patterns should not be influenced by forest succession. Water samples were collected biweekly from four streams in, and near, the old-growth forest watershed of the Bowl Research Natural Area in the White Mountains of New Hampshire from May 1973 through October 1974, and from June 1994 through June 1997.
Vol. 30; no. 8; pp. 1206-1213

Martin, L.R.; Hill, M.W.; Tai, A.F.; Good, T.W.
The iron catalyzed oxidation of sulfur(IV) in aqueous solution: Differing effects of organics at high and low pH. Journal of Geophysical Research; Feb 20 1991.
Description: The authors have studied the oxidation of sulfur dioxide by dissolved oxygen in highly dilute solutions with a new differential optical absorption technique. The authors measured the rate of oxidation catalyzed by iron (III) over a wide range of pH, ionic strength, and in the presence of various organic materials. The studies indicate that noncomplexing organic molecules are highly inhibiting at high pH values of 5 and above and are not inhibiting at low pH values of 3 and below.
Vol. 962; pp. 3085-3097

Martin, P.J.S; Clark, J.M; Edman, J.D.
Preliminary study of synergism of acid rain and diflubenzuron
Bulletin of Environmental Contamination and Toxicology; Jun 30 1995
Description: Diflubenzuron[1] (Dimilin{reg_sign}) was used on over 7 million acres in the U.S. in 1990 to control forest pests, particularly the gypsy moth. This chitin synthesis inhibitor affects insects and other anthropods. It is a restricted use pesticide due to its nontarget effects on aquatic macroinvertebrates. The effects of a single aerial application on nontarget aquatic macroinvertebrate communities were reviewed by Eisler (1992). Crustacea and immature insects (especially the true flies, mosquitoes, midges and black flies) are the most sensitive nontarget aquatic organisms to diflubenzuron. Diflubenzuron, N-[[4-(chlorophenyl)amino]carbonyl]-2,6-difluorobenzami de, is not the only mortality factor aquatic organisms face from human pollution.
Vol. 54; no. 6; pp. 833-836

Martin, R.B.
Aluminum: A neurotoxic product of acid rain. Accounts of Chemical Research; Jul

31 1994. Description: Two separate but converging concerns have resulted in an upsurge in research on aluminum ion in the past 15 years. Acid rain releases Al(III) from soils into fresh waters, where it is for the first time accessible to living organisms. Though long considered benign, Al(III) has recently been found to cause bone and neurological disorders, while its role in Alzheimer's disease remains uncertain. The greater availability of Al(III), coupled with its demonstrated harmful effects, challenges chemists to describe its chemistry and biochemistry. Many interactions of Al(III) have been described, but several questions remain unsolved. Vol. 27; no. 7; pp. 204-210

Massey, D.D
Coal: An industry in transition
Electric Perspectives; Jan 31 1995
Description: This article presents a detailed review of the coal industry. Data presented and analyzed for a 12-year period include coal prices, coal mining productivity, and number and size of coal mines. Electric industry issues, such as consolidation and deregulation, are described in terms of their impact on the coal industry. Merger and acquisition activities and acid rain legislation are also outlined.
Vol. 20; no. 1; pp. 30-42

Massucci, M; Clegg, S.L; Brimblecombe, P.
Equilibrium vapor pressure of H_2O above aqueous H_2SO_4 at low temperature
Journal of Chemical and Engineering Data; Jul 31 1996. Description: Equilibrium partial pressures of water ($p(H_2O)$) over solutions of (5.054 to 26.18) mol/kg H_2SO_4 from (31.4.2 to 196.5) K have been determined using capacitance manometers in order to (1) obtain data for compositions relevant to the stratospheric aerosol and (2) extend and test present knowledge of the thermodynamic properties of aqueous H_2SO_4 at low temperature. Current approaches to calculating $p(H_2O)$ from existing thermodynamic data are critically examined and compared to measurements made in this study and others in the literature. Vol. 41; no. 4; pp. 765-778

Matzner, E.; Meiwes, K.J.
Long-term development of elementa fluxes with bulk precipitation and throughfall in two German forests. Journal of Environmental Quality; Jan 31 1994. Description: Long-term monitoring of element fluxes with bulk precipitation and throughfall in two forest ecosystems of the German Soiling area revealed substantial changes in the chemical climate of these ecosystems. From 1966 to 1990, fluxes of NO_3 and NH_3 in throughfall increased by 40 to 60%. In contrast, SO_4, H^+ and Ca fluxes decreased by 30 to 60%.
Vol. 23; no. 1; pp. 162-166

McCally, M.; Cassel, C.K.
Medical responsibility and global environmental change. Annals of Internal Medicine; Sep 15 1990. Description: Global environmental change threatens the habitability of the planet and the health of its inhabitants. Toxic pollution of air and water, acid rain, destruction of stratospheric ozone, waste, species extinction and, potentially, global warming are produced by the growing numbers and activities of human beings. Progression of these environmental changes could lead to unprecedented human suffering. Physicians can treat persons experiencing the consequences of environmental change but cannot individually prevent the cause of their suffering. Physicians have information and expertise about environmental change that can contribute to its slowing or prevention. Work to prevent global environmental change is consistent with the social responsibility of physicians and other health professionals. 44 references. Vol. 1136; pp. 467-473

McCormick, L.H.
Variation in mineral content of red maple sap across an atmospheric deposition gradient. Communications in Soil Science and Plant Analysis; Jan 31 1997. Description: Xylem sap was collected from red maple (Acer rubrum L.) trees during the spring of 1988 and 1989 at seven forest sites along an atmospheric deposition gradient in north central Pennsylvania and analyzed for pH and twelve mineral

constituents. The objectives of the study were to examine the sources and patterns of variation in red maple sap chemistry across an atmospheric deposition gradient and to assess the feasibility of using sap analysis as an indicator of nutrient bioavailability. For most sap constituents, there was considerable spatial and temporal variation in concentration. Vol. 28; no. 3-5; pp. 365-379

McCormick, W.T. Jr.
Natural gas as a 'natural' solution Public Utilities Fortnightly; May 15 1991 Description: This article promotes natural gas use as a means to cut US dependence on imported oil by some 28 percent over the next ten years, while improving energy efficiency and solving a portion of the global warming and acid rain problems. Topics of discussion include fuel substitution, the Clean Air Act, natural gas capacity and distribution, and natural gas exploration. Vol. 12710; pp. 94-95

McDonald, A.
Combating acid deposition and climate change - priorities for Asia. Environment; Jul 01 1999. Description: This paper is about two types of interactions that are relevant to efforts to limit both acid rain and global warming. The first are the interactions between sulphur and carbon emissions in producing the impacts that ultimately matter, such as the impacts on agriculture in Asia. The second are behavioural interactions, that is, the rational responses to policies targeting one problem that make the other more difficult to solve. Vol. 41; no. 3; pp. 4-11,34-41

McHenry, J.N.; Dennis, R.L.
The relative importance of oxidation pathways and clouds to atmospheric ambient sulfate production as predicted by the regional acid deposition model. Journal of Applied Meteorology; Jul 31 1994. Description: The development and use of a version of the Regional Acid Deposition Model/Engineering Model (RADM/EM) called the Comprehensive Sulfate Tracking Model (COMSTM) is reported. The COMSTM is used to diagnose the relative contributions of each sulfate production pathway to the total atmospheric ambient sulfate predicted by RADM. Thirty meteorological cases are used to aggregate the results into annual estimates. For the operational RADM (denoted RADM 2.6), nonprecipitating cloud production of ambient sulfate dominates over precipitating cloud production, and the hydrogen peroxide pathway dominates over four other aqueous formation routes. Gas-phase production of sulfate contributes less than 40% of the ambient budget. Vol. 33; no. 7; pp. 890-90

McIlvaine, R.W.
Pollution solutions in Wonderland. Environmental Solutions; Aug 31 1994. Description: The Environmental Protection Agency (EPA) calculates coal-fired boiler fine-dust emissions at 100,000 tons per year (tpy). However, the actual amount is well in excess of 700,000 tpy. It is as if EPA calculated the national average gas mileage by using auto dealers' miles-per-gallon claims for new subcompact cars, but ignored the 20-year-old cars still on the road. Vol. 78; pp. 46-47

McIntosh, H.
Catching up on the Clean Air Act [news] Environmental Health Perspectives; Aug 31 1993. Description: No abstract available. Vol. 1013; pp. 226-229

McKenna, M.A.; Hille-Salgueiro, M.; Musselman, R.C.
Effects of acid precipitation on reproduction in alpine plant species. American Journal of Botany; Jan 31 1990; pp. 59. Description: A series of experiments were designed to determine the impact of acid rain on plant reproductive processes, a critical component of a species life history. Research was carried out in herbaceous alpine communities at the USDA (United States Department of Agriculture) Forest Service Glacier Lakes Ecosystem Experiments Site in the Snowy Mts. of Wyoming. A range of species were surveyed to monitor the sensitivity of pollen to acidification during germination

and growth, and all species demonstrated reduced in vitro pollen germination in acidified media.

McLaughlin, J.W; Reed, D.D; Jurgensen, M.F. Foliar amino acid accumulation as an indicator of ecosystem stress for first-year sugar maple seedlings. Journal of Environmental Quality; Jan 31 1994. Description: Accumulation of certain plant foliar amino acids (arginine, glutamine, and proline) can be used as indicators of anthropogenic and natural stressors, such as atmospheric deposition and mineral nutritional imbalances, which result in decreased plant growth. In this study a number of factors were evaluated to assess the use of foliar amino acid accumulation as indicators of sugar maple seedling stress at two sugar maple dominated forests in Michigan. These factors were: (1) first-year sugar maple (Acer saccharum Marshall) seedling growth, (2) N and P nutrition, (3) soluble foliar and root total amino acid concentrations, and (4) concentrations of foliar arginine, glutamine, and proline.
Vol. 23; no. 1; pp. 154-161

McLaughlin, S.B; Edwards, N.T; Hanson, P.J. Growth responses of 53 open-pollinated loblolly pine families to ozone and acid rain. Journal of Environmental Quality; Mar 31 1994. Description: Field exposures of 9950 containerized 12-wk-old loblolly pine (Pinustaeda L.) seedlings representing 53 commercially important, open-pollinated families were conducted to evaluate individual and interactive effects of acid rain and O_3 on growth response. A 36-plot field research facility comprised of 33 open-top chambers and three open plots was used to test effects of five O_3 levels that included ambient (A) and seasonally integrated levels that were 0.53, 1.10, 1.58, or 2.15 times ambient. Individual effects of three levels of simulated acid rain (pH 3.3, 4.5, and 5.2) as well as their interaction with O_3 at 0.53A, 1.58A, and 2.15A levels were also included.
Vol. 23; no. 2; pp. 247-257

McRae, D.J.; Duchesne, L.C.; Freedman, B.; Lynham, T.J.; Woodley, S. Comparisons between wildfire and forest harvesting and their implications in forest management. Environmental Reviews; Dec 01 2001. Description: Emulation silviculture is the use of silvicultural techniques that try to imitate natural disturbances such as wildfire. Emulation silviculture is becoming increasingly popular in Canada because it may help circumvent the political and environmental difficulties associated with intensive forest harvesting practices. In this review we summarize empirical evidence that illustrates disparities between forest harvesting and wildfire.
Vol. 9; no. 4; pp. 223-260

Meilinger, S.K; Koop, T; Luo, B.P. Size-dependent stratospheric droplet composition in lee wave temperature fluctuations and their potential role in PSC freezing. Geophysical Research Letters; Nov 15 1995. Description: This report discusses the effect of rapid temperature fluctuations on stratopheric droplet size composition. Stoichiometry of these droplets suggest that this could be a cause for the formation of frozen polar stratospheric clouds of type-Ia.
Vol. 22; no. 22; pp. 3031-3034

Menz, F.C
Transborder emissions trading between Canada and the United States
Natural Resources Journal; Jan 31 1995
Description: Despite the fact that legislation to control acid rain is now in place in Both Canada and the United States, it would still be in both countries' best interest to deal explicitly with transboundary flows of sulfur dioxide. Dealing directly with transboundary acid rain as a reciprocal externality suggests a commitment to binational management that does not currently exist and allows for more cost-effective pollution control by encouraging abatement at lower-cost sources irrespective of national borders. A transborder emissions trading program for northeastern North America, an area encompassing eastern Ontario,

southwestern Quebec, northern New York, and northern New England, is proposed. Vol. 35; no. 4; pp. 803-819

Menzel, R.G.
Long-term field research on water and environmental quality. Agronomy Journal; Jan 31 1991. Description: Maintaining the quality of soil, water, and air presently has a high national priority, and is particularly important for agriculture. The deterioration of environmental quality, in most respects, has proceeded almost imperceptibly. Long-term research is required to study effects that change slowly with time, such as contamination of groundwater, or that depend upon infrequent weather events, such as water erosion of soil.
Vol. 831; pp. 44-49

Miller, K.W.; Cole, M.A.; Banwart, W.L.
Microbial populations in an agronomically managed mollisol treated with simulated acid rain. Journal of Environmental Quality; Jan 31 1991. Description: A fertile well-buffered mollisol (Flanagan silt loam, fine montmorillonitic mesic Aquic Arguidoll) under cultivation with corn (Zea mays L.) and soybean (Glycine max (L.) Merr.) was subjected to simulated rain of pH 5.6, 4.2, and 3.0, while moisture-activated rain exclusion shelters provided protection from ambient rain. Soil was sampled to a depth of 3 cm on four dates throughout the 1985 growing season. The following microorganisms were enumerated by plate counts or most probable number: general heterotrophic bacteria, general soil fungi, free-living N-fixing bacteria, fluorescent pseudomonads, autotrophic ammonium-oxidizing, nitrite-oxidizing, and thio-sulfate-oxidizing bacteria. Vol. 204; pp. 845-849

Miller, R.W
The outlook for US nuclear power NUEXCO. Monthly Report to the Nuclear Industry; Dec 31 1988. Description: The last firm US order for a commercial nuclear power plant was placed in 1975. By the time the next order is placed around the year 2000, a full quarter-century will have elapsed between orders. A few prototype or R&D reactors could be ordered in the early 1990s, but uranium will not offer any significant competition to coal in the choice of fuels for generating capacity orders placed in the rest of this century. Coal's primary competition will come from gas. At first, economic factors caused the half in US orders, but nuclear power's lack of public acceptance is the most important reason that the halt has continued. No. 244; pp. 25-28

Miller, Steven.L.; McClean, Therese.M.; Stanton, Nancy.L.; Williams, Stephen.E.
Mycorrhization, physiognomy, and first-year survivability Park. Canadian Journal of Forest Research; Jan 01 1998. Description: Ectomycorrhiza formation, survivability, and physiognomic characteristics were assessed for conifer seedlings encountered 1 and 2 years postfire in the Huck burn site near Grand Teton National Park. *Pinus contorta* Dougl. ex Loud. germinated and was abundant throughout the first growing season. *Abies lasiocarpa* (Hook.) Nutt. germinated during May and June but was rarely encountered by September.
Vol. 28; no. 1; pp. 115-122

Mitnick, S.; Brown, A.; McKinney, K.; White, J.; Henry, J.
Allowance trading today and tomorrow: When will it really get started. Electricity Journal; Oct 31 1992. Description: This article is the product of a conference call among five individuals who have been active in formulation and implementation of the acid rain provisions of the 1990 Clean Air Act amendments. They take stock of where we are, where we are headed, and what it will take to get there. Vol. 58; pp. 62-72

Molau, Ulf.; Larsson, Eva-Lena.
Seed rain and seed bank along an alpine altitudinal gradient in Swedish Lapland. Canadian Journal of Botany; Jun 01 2000 Description: We studied the seed flux, including seed rain and seed bank (germinable and total), at twelve sites along an altitudinal gradient in the Abisko area in northernmost Swedish Lapland

during a period of 3 years with contrasting summer climates. The study sites were evenly spaced in altitude from the timberline at 700 m above sea level to the highest peaks in the area (1560 m). A subalpine birch forest site was included for comparison. Vol. 78; no. 6; pp. 728-747

Molburg, J.C.
The utility industry response to Title IV: generation mix, fuel choice, emissions and costs. Air and Waste; Feb 28 1993. Description: The Clean Air Act Amendments of 1990 incorporate, for the first time, provisions aimed specifically at the control of acid rain. These provisions restrict emissions of sulfur dioxide and oxides of nitrogen from electric power generating stations. The restrictions on sulfur dioxide take the form of an overall cap on the aggregate emissions from major generating plants, allowing substantial flexibility in the industry's response to those restrictions. Vol. 432; pp. 180-186

Mookerjee, G.; Dean, A.T.
Continuous emission monitoring system and its compliance with federal clean air act amendments. Proceedings of the American Power Conference; Jan 31 1992 Description: The CEM System, utilizing a microprocessor based Data Acquisition System (DAS), stores emission data and generates reports which comply with EPA's emission data reporting requirements. Detroit Edison has nine fossil-fueled power plants and is planning to install twenty CEM systems, one per each stack. The company is presently in the process of installing tow such system sat Monroe Power Plant. This paper describes these two CEM systems.
Vol. 541; pp. 700-704

Moore, A.
Putting the prices together - The Clean Air Act puzzle. Waste Age; Mar 31 1991 Description: This paper reports that after 12 years of heated debate, closed-door negotiations, and massive opposition by business and industry, the Clean Air Act (CAA) has finally been revamped. NSWMA participated actively and extensively in the CAA reauthorization process to ensure that unnecessary compliance requirements were not imposed on the waste management industry. The new law embodies five critical objectives that the industry sought and received. The new law consists of 11 titles. Seven titles will potentially impact the waste management industry. Titles I-IV address specific sources of air pollution including smog (ozone depleter) and carbon monoxide, motor vehicles, acid rain form utility plants, and unregulated hazardous air pollutants emitted by both large and small facilities. Vol. 223; pp. 90-105

Moore, T
New water regulations on the horizon Electric Perspectives; Mar 31 1997. Description: Many electric utilities are still feeling the economic impact that recent major environmental initiatives--such as those related to the control of emissions that cause acid rain and to the installation of continuous emissions monitors--have had on fossil power generation. Now another wave of legislative and regulatory activity, compliance effort, and potential cost is on the horizon. The first major round of revisions to the nation's clean water laws in a decade is being planned in Congress, at the US Environmental Protection Agency (EPA), and by regional commissions and individual states.
Vol. 22; no. 2; pp. 20-26

Morales-Baquero, Rafael.; Presentación Carrillo; Reche, Isabel.; Pedro Sánchez-Castillo. Nitrogen-phosphorus relationship in high mountain lakes: effects of the size of catchment basins. Canadian Journal of Fisheries and Aquatic Sciences; Oct 01 1999. Description: We analyzed the changes in epilimnetic total nitrogen (TN), total phosphorus (TP), dissolved inorganic nitrogen (DIN), and soluble reactive phosphorus (SRP) in 31 small high-mountain lakes in the Sierra Nevada (Spain) during an annual cycle, just after the spring thaw, and in the middle of the growing season. Chlorophyll *a*, TN, and TP increased, whereas the TN:TP ratio fell substantially between the two periods,

reaching values generally between 25 and 10 (by weight). On the contrary, DIN, SRP, and DIN:SRP ratios were similar for both periods in each lake.
Vol. 56; no. 10; pp. 1809-1817

Morgan, M.
New uses for ORNL's ultrasensitive mass spectrometer. Oak Ridge National Laboratory Review; Jan 31 1993. Description: Oak Ridge National Laboratory has an ultrasensitive instrument that can reduce the cost of monitoring workers for radiation exposure, determine concentrations of trace elements in tree cores to assess the effects of acid rain on soil chemistry and tree growth, and even identify counterfeit bolts that need to be replaced. The inductively coupled plasma mass spectrometer (ICP-MS). May someday demonstrate its value is in the DOE-required monitoring of workers exposed to radioactive materials such as uranium. Vol. 261; pp. 16-21

Morgan, M.D.
Streams in the New Jersey Pinelands directly reflect changes in atmospheric deposition chemistry. Journal of Environmental Quality; Jan 31 1990. Description: The major ion chemistry of precipitation and four undisturbed streams was measured in the New Jersey Pinelands (Pine Barrens) in 1970-1972 and 1984-1988. Over the interval, the sea salt corrected concentration of Ca and SO_4 significantly declined in precipitation and all streams. Analytical problems prevented analysis of the change in precipitation H, but H significantly declined in all streams. The change in precipitation and stream water Mg and K was less consistent, but was generally toward lower concentrations in 1984-1988. Vol. 192; pp. 296-302

Morgan, M.D.
Sulfur pool sizes and stable isotope ratios in HUMEX peat before and immediately after the onset of acidification. Environment International; Jan 31 1992. Description: Total nonsulfate sulfur (TS), reduced inorganic sulfur (RIS), ester sulfate (ES) and carbon bonded sulfur (CBS) pool sizes, and TS and RIS stable sulfur isotopes were examined in peat cores from control and experimental sites within the Skjervatjern catchment before and 8.5 months after the onset of acidification. The total amount of excess sulfate added to the catchment up to this point was less than expected due to mechanical problems. There have been no observable effects on sulfur cycling within the peat soils. Vol. 186; pp. 545-553

Moyad, A.
US acid-rain policy, in court vs. in congress: Heads or tails. Water Environment Technology; Jul 31 1990. Description: This article discusses the official US position on acid rain, which is difficult to ascertain based on legislation and court decisions. Vol. 27; pp. 30-31

Moynihan, D.P.
Acid precipitation and scientific fallout. Forum for Applied Research and Public Policy; Jan 31 1993. Description: U.S. Senator Daniel Patrick Moynihan from New York believes the nation should set its environmental policies based upon knowledge of the costs and benefits of such policies. For that reason, he defends the National Acid Precipitation Assessment Program (NAPAP), despite the program's lack of influence on recent clean-air legislation. "Environmentalism is nothing if not an ethic of responsibility," Senator Moynihan writes, "and our first responsibility is to the facts-facts about costs and facts about benefits."
Vol. 82; pp. 61-65

Mudur, G
Monsoon shrinks with aerosol models. Science (Washington, D.C.); Dec 22 1995. Description: Sulfur compounds in the atmosphere from burning fossilfuels have long been implicated in acid rain and smog, but climate modelers have only recently included them along with greenhouse gases as major determinants of global change. This paper summarizes work of German modelers who showed just how important the aerosols that form from the sulfur compounds may be in

affecting regional climates. Among the model findings is the projection of a decrease in monsoons in India due to the increase in aerosol levels.
Vol. 270; no. 5244; pp. 1922

Muenster, U.
Microbial extracellular enzyme activities in HUMEX Lake Skjervatjern
Environment International; Jan 31 1992
Description: Two microbial extracellular enzyme activities (MEEA) were studied in HUMEX Lake Skjervatjern: acid phosphatase (APHA) and leucine aminopeptidase (LeuAMPA). Both enzyme activities varied in the vertical and horizontal scale in both lake sites. APHA varied in the acidfied Basin A between 945-1706 nmol L^{-1} h^{-1} and LeuAMPA between 3.7-25 nmol L^{-1} h^{-1}. Both MEEA reached maxima in 0.5 m depth. In the control site (Basin B), APHA was lower by a factor of two, and varied between 156-669 nmol L^{-1} h^{-1}. LeuAMPA reached similar values as in Basin A and varied between 7.8- 34.8 nmol L^{-1} h^{-1}. Maxima of APHA were found in the upper layer (0-2 m), while LeuAMPA had only one distinct maxima at 2-2.5 m depth.
Vol. 186; pp. 637-647

Mulder, J.; Christophersen, N.; Hauhs, M.; Vogt, R.D.; Andersen, S.; Andersen, D.O.
Water flow paths and hydrochemical controls in the Birkenes catchment as inferred from a rainstorm high in seasalts
Water Resources Research; Apr 30 1990
Description: At Birkenes, a small forested catchment with acidic soils in southernmost Norway, acid rain has resulted in high stream water H^+, Al, and SO_4 concentrations. Recent studies have revealed the complexity of the Al chemistry in Birkenes stream water, as inorganic Al is not regulated by one single solubility control. It has been hypothesized that this is due to the dynamic nature of water flow paths and the different Al solubilities in surface soils and subsoils.
Vol. 264; pp. 611-622

Munn, Ted; Timmerman, Peter; Whyte, Anne
Emerging Environmental Issues. Bulletin of the American Meteorological Society; Jul 01 2000. Description: Emerging environmental issues are issues that may someday be of concern but that have not yet been generally recognized. A review of such issues that have occurred over the last 50 years reveals that many of them have erupted rather suddenly (e.g., stratospheric ozone depletion, acid rain). However, some issues were recognized long ago by the scientific community (e.g., land degradation, overconsumption of freshwater), but for economic or other reasons governments have refused to act.
Vol. 81; no. 7; pp. 1603-1609

Myers, N.
The question of linkages in environment and development. Bioscience; May 31 1993. Description: To make the world more manageable, humans have split it up into disciplinary components such as nations, communities, economic sectors, ecological zones, and scientific disciplines. However, the preoccupation with a certain sector often means the larger perspective is lost. Dynamic interactions between the sectors are as important as the sectors themselves. Vol. 435; pp. 302-310

Mykkelbost, T.C; Rogt, R.D; Seip, H.M.
Organic carbon fractionation applied to lake- and soilwater at the Humex Site
Environment International; Jan 31 1995
Description: Effects of artificial addition of acid precipitation on dissolved organic carbon (DOC) were studied at the Humic Lake Acidification Experiment (HUMEX) catchment on the western coast of Norway. Samples of soil water and lake water from a treatment and reference side were separated into different DOC fractions by the use of size fractionation and column chromatography with XAD-8 and ion-exchange resins. The main DOC components are hydrophobic and hydrophilic acids (HPO-A and HPI-A). While the amount and composition of DOC were stable down in a histosol profile, the amount decreased and composition changed down in a podzol profile. Vol. 21; no. 6; pp. 849-859

Nagelhout, M.
  State Regulatory responses to acid rain: Implications for electric utility operations
  Public Utilities Fortnightly; Mar 01 1990
  Description: This article discusses the state regulatory responses to acid rain legislation and how this will affect electric utility operations. Topics discusses include planning and fuel procurement practices, least-cost planning, long-term supply contracts, fuel mix, cogeneration and small power production, qualifying facility contracts, avoided costs, environmental impact... Vol. 1255; pp. 43-46

Nan, G.D.
  Fuels' social costs: Evidence from all electric-generating firms. Journal of Energy and Development; Jan 31 1993.
  Description: The US Clean Air Act of 1970, which amended the Air Quality Act of 1967, was itself amended in 1977 and 1990. The act and its amendments have significantly impacted on the power industry. Some research projects have studied these effects intensively, for example, T.J. Stanton and F.M. Gollop and M.J. Roberts. The Clean Air Act Amendments of 1990 (CAAA) established a totally new control scheme for addressing the acid rain problem in Title IV, which focused on power plants emissions of sulfur dioxide (SO_2) and nitrogen oxide (NO_x). It is believed that the CAAA will have some impacts on affected utilities' fuel choices. This research attempts to provide some quantitative information concerning the possible impacts. Vol. 18; no. 2; pp. 233-241

NAPAP (National Acid Precipitation Assessment Program) results on acid rain
  Journal of Forestry; Jun 30 1990
  Description: The National Acid Precipitation Assessment Program (NAPAP) was mandated by Congress in 1980 to study the effects of acid rain. The results of 10 years of research on the effect of acid deposition and ozone on forests, particularly high elevation spruce and fir, southern pines, eastern hardwoods and western conifers, will be published this year. Vol. 886; pp. 10

NARUC: Regulation at the crossroads
  Public Utilities Fortnightly; Jan 04 1990
  Description: This article reports on the review of changes during the last decade of the regulation of industries by the federal government. The state commissioners that make up the National Association of Regulatory Utility Commissioners discuss the changing regulatory climate for a competitive industry, changes in the federal- state relationship, proposed changes in jurisdictional boundaries, acid rain legislation and emissions trading, and Senator J. Bennet Johnson's amendment to the Public Utilities Holding Company Act. Vol. 1251; pp. 37-38

Nelson, G.L.
  Taxing sulfur dioxide emission allowances Fortnightly; Sep 15 1993. Description: The acid rain control program authorized by Title IV of the Clean Air Act Amendments of 1990 (CAAA) was designed to reduce the adverse effects of acid rain by limiting emissions of sulfur dioxide (SO_2) into the atmosphere. The program is a complex scheme involving the issuance, consumption, holding, and trading of emission allowances for SO_2. Not surprisingly, electric utilities will face federal income tax issues in connection with the program. Vol. 13117; pp. 46-47

Neufeld, R.D.; Vallejo, L.E.; Hu, W.; Latona, M.; Carson, C.; Kelly, C.
  Properties of high fly ash content cellular concrete. Journal of Energy Engineering; Apr 30 1994. Description: High fly ash content autoclaved cellular concrete is produced by adding calibrated quantities of aluminum powder to a mixture of fly ash (60% wt/wt), cement, and water. The foamed product is hardened in an autoclave with pressurized steam at about 180 C. Block material for samples tested originated from a mobile pilot plant that toured sites of United States-based electric utilities. Vol. 1201; pp. 35-48

Neuman, L.; Houpis, J.; Anderson, P.
  Among clone comparison of the response of photosynthetic pigments in mature branches of Pinus ponderosa to long-term

exposure of ozone and acid rain. Plant Physiology, Supplement; May 31 1990. Description: Three clone families of mature grafted scions are to be exposed to ambient, charcoal filtered, and ozone-enriched air both separately and in conjunction with two concentrations of acid precipitation (3.0 pH, 5.1 pH) using branch exposure chambers (BECs). A total of 72 branches are being studied from 18 trees. Ambient ozone will be continuously monitored and the elevated ozone treatment will double this value (2x ambient). Acid rain events will occur weekly from February through May to provide total precipitation amounts similar to the annual rainfall occurring in the central Sierra-Nevada of California. Needle samples were analyzed spectrophotometrically to determine the absorbance of chlorophyll a and b and carotenoids at 662 nm, 645 nm, and 470 nm, respectively. Pigment concentrations were calculated for surface area (mg/cm^2) and compared among individuals and among clones. Pre-exposure concentrations of chlorophyll a and carotenoids were found to differ significantly (=0.05) among clones. Vol. 931; pp. 101

Neuman, L.; Houpis, J.; Anderson, P.
Trends in Pinus ponderosa foliar pigment concentration due to chronic exposure of ozone and acid rain. Plant Physiology, Supplement; May 31 1991. Description: To determine the effects of ozone and acid rain on mature Ponderosa pine trees, Lawrence Livermore National Lab. has collaborated with University of California Berkeley, University of California Davis, California State University Chico, and the US Forest Service at the latter's Chico Tree Improvement Center. Foliar tissue from mature grafted scions of Pinus ponderosa were exposed to two times ambient ozone for ten months and to acid rain (3.0 pH) weekly for 10 weeks using branch exposure chambers. Vol. 961; pp. 172

New form of calcium carbonate improves SO_2 removal from boilers. Air Pollution Consultant; Nov 30 1996. Description: As acid rain control regulations take effect, some utility companies are considering or have installed flue gas desulfurization (FGD) systems using lime-based sorbents. With one type of FGD system, called furnace sorbent injection (FSI), sorbents are injected directly into the combustion chamber of a coal-fired boiler. Such systems have proven effective at reducing sulfur dioxide (SO_2) emissions from some types of boilers, especially those that operate at relatively low temperatures. However, FSI systems have generally not performed well with most types of conventional boilers. A New York company has patented a new FGD sorbent called thermally active marble (TAM). TAMs tend to fracture and expose new reaction surfaces - much like ice cubes in hot water. Vol. 6; no. 7; pp. 1.23-1.24

Ng, W.S.; Buttorff, L.A.; Easley, M.
System planning in the 1990s: New choices, new criteria, new solutions. Proceedings of the American Power Conference; Jan 31 1991. Description: The electric utility resource planning process has evolved from evaluating the economics and timing of future unit installations to a much more sophisticated analysis of all practical alternative resources, including those that reduce electricity consumption. The term resources how connotes a wide range of demand- side and supply-side options. Traditional resource planning was essentially confined to preparing load forecasts and matching load with new generating capacity to achieve a pre-set reserve margin or a reliability index. Vol. 53; pp. 226-230

Nikolaidis, N.P.; Ecsedy, C.; Olem, H.; Nikolaidis, V.S.
Acidic deposition and global climate change. Research Journal of the Water Pollution Control Federation; Jun 30 1990. Description: A literature is presented which examines the research published on understanding ecosystem acidification and the effects of acidic deposition on freshwaters. Topics of discussion include the following: acidic deposition; regional assessments; atmospheric deposition and transport; aquatic effects; mathematical

Nikolaidis, N.P.; Muller, P.K.; Schnoor, J.L.; Hu, Hsienlun
Modeling the hydrogeochemical response of a stream to acid deposition using the enhanced trickle-down model. Research Journal of the Water Pollution Control Federation; Jan 31 1991. Description: The enhanced trickle-down model was applied to White Oak Run, a second-order stream located in Shenandoah National Park in Virginia. Calibration of the model was performed using 5 years of field data. Simulation results indicate that 29% of the incoming sulfate is retained in the upper soil horizons and 22% is retained in the lower soil horizons. The alkalinity concentrations in the stream exhibited a seasonal variation, with increased values during low flow and decreased values during high flow. Ion exchange in the upper soil neutralized 39% of the incoming acidity and weathering accounted for 14% of the neutralization in the lower soil horizons. Vol. 633; pp. 220-227

Nomura, Toshikazu; Ehara, Yoshiyasu; Kishida, Haruo; Ito, Tairo
A study of NO removal by packed-beads discharge reactor. IEEE Transactions on Industry Applications; Dec 01 1999. Description: Recently, air pollution has become a serious problem; photochemical smog and acid rain are typical phenomena. NO_x is a serious air pollutant and a toxic gas. In spite of an attempt to reduce the amount of NO_x emitted, the density of NO_x in the atmosphere has remained on a stable level, or even become worse.
Vol. 35; no. 6; pp. 1318-1322

Nordhaus, R.R.
Acid rain compliance: Coordination of state and federal regulation
Energy Law Journal; Jan 31 1992
Description: The Clean Air Act (CAA) Amendments of 1990 impose new controls on emissions by electric utilities of the two major precursors of acid rain: sulfur dioxide (SO_2) and oxides of nitrogen (NO_x). Utilities, and the utility holding company systems and power pools of which they are members, will be subject to extensive and costly compliance obligations under the new statute. Most of these utilities, utility systems, and power pools are regulated by more than one utility regulatory authority. Some utilities are regulated by several states, some by a single state and by the Federal Energy Regulatory Commission (FERC), and some by multiple states, by the FERC, and by the Securities and Exchange Commission (SEC). Vol. 132; pp. 341-357

N'soukpoe-Kossi, C.N.; Trottier, C.; Achi, C.A.; Charlebois, D.; Leblanc, R.M.
Effects of acid watering of the soil on the photosynthetic activity, growth, and foliar pigments of sugar maple saplings
Journal of Environmental Science and Health, Part A: Environmental Science and Engineering; Jan 31 1992. Description: Photoacoustic spectroscopy has been used to monitor the photosynthetic activity of the leaves of sugar maple saplings treated by watering the soil with simulated acid rain at different pH levels (2.5, 3.0, 3.5, 4.0, 4.5, 5.0 and 5.6). We also measured the relative growth rate of the plants and the pigment content of leaves. The results indicated an ambivalent effect of acidity:
Vol. 273; pp. 863-877

Odnevall, I; Leygraf, C.
Atmospheric corrosion of copper in a rural atmosphere. Journal of the Electrochemical Society; Nov 30 1995. Description: A field exposure program has been implemented in a rural atmosphere with the primary aim of studying the corrosion mechanisms of sheltered copper. Emphasis has been placed upon the initial corrosion behavior after days, weeks, and months. Two starting dates, one in October and one in April, with different environmental characteristics have been studied. By using a multianalytical approach combined with environmental characterizations, a picture of the corrosion processes has emerged. Initially, a film of cuprite (Cu_2O) is formed. Vol. 142; no. 11; pp. 3682-3689

Olem, H.
> Ten-year study on acid precipitation nears conclusion. Water Environment amp Technology; Apr 30 1990. Description: Results from the National Acid Precipitation Assessment Program (NAPAP) are discussed. Final results are contained in 26 state of the science reports. Seven of the reports provide information on acid rain and aquatic ecosystems. They describe the current state of acidic surface waters, watershed processes affecting surface water chemistry, historical evidence for surface water acidification, methods for forecasting future changes, and the response of acidic surface water to liming. Vol. 24; pp. 11-13

Orchard, D.
> The Green Plan: A national challenge for Canada. Journal of the Air and Waste Management Association; May 31 1991. Description: Canada's new Green Plan charts a course for dealing with environmental issues that will require an unflinching commitment to changes in thought, decision-making and action by all sectors of Canadian society. The five-year, $3-billion Plan lists 120 specific initiatives in eight areas of activity. It will enlist the participation of some 40 federal departments and agencies, alone, or in creative partnerships with industry, the academic and scientific community, and citizens' groups. Vol. 415; pp. 268-271

Oxburgh, R.; Drever, J.I.; Sun, Yan-Ting
> Mechanism of plagioclase dissolution in acid solution at 25°C. Geochimica et Cosmochimica Acta; Jan 31 1994. Description: The dissolution kinetics of three plagioclase feldspars (An_{13}, An_{46}, and An_{76}) were studied in flow-through reactors over the pH range 3-7. In accordance with the surface complexation model, dissolution rate was described by the equation Rate = $[chi]_a A \exp(E_a/kT) (C^5_H)^n$, where (C^5_H) is the concentration of protonated surface sites, $[chi]_a$ is the mole fraction of these that are activated, A is the Arrhenius pre-exponential factor, and E_a is the activation energy for the reaction. Vol. 582; pp. 661-669

P.C., Van Deusen,; Reams, G.A.; Cook, E.R.
> Possible red spruce decline: Contributions of tree-ring analysis. Journal of Forestry; Jan 31 1991. Description: Debate continues about the cause of apparent unprecedented decreases in ring width at all elevations, and increasing levels of mortality at high elevations, in red spruce (Picea rubens) stands in the northeastern United States. These growth and mortality trends are often used as evidence of red spruce decline, but the possibility remains that they may be occurring naturally. Two hypotheses are being used to explain the causes of red spruce growth reduction across its range and increased levels of standing dead at some high-elevation sites. Vol. 891; pp. 20-24

Padmanabha, A.P.; Olem, H.
> The new Clean Air Act. Water Environment amp Technology; May 31 1991. Description: This article is a title by title review of the new Clean Air Act and how it affects water quality and wastewater treatment. The bill provides for restoring and protecting lakes and rivers by reducing acid-rain-causing emissions and toxics from nonpoint-source runoff. Topics covered include urban smog, mobile sources, air toxics, acid rain, permits, ozone-depleting chemicals, enforcement, and the law's socio-economic impacts. Vol. 35; pp. 40-46

Palola, E.
> Wood opportunities. Independent Energy; Jan 31 1991. Description: This article examines the impacts of the new Clean Air Act amendments on wood-fired power plant projects. The topics of the article include identifying areas of the amendments where there are potential constraints and opportunities for wood-fired power plant projects, geography, waste wood scenarios, sorting out the rules, acid rain advantages, a boost for renewables, and resources for further investigation. Vol. 216; pp. 49-52

Parkinson, G.; Ondrey, G.; Moore, S.
> Sulfur production continues to rise Chemical Engineering (New York); Jun 30

1994. Description: Sulfur is one of the world's most-popular commodities. It has another distinctive feature: most of it is produced from the effluent of chemical process plants. A lot more sulfur will have similar origins in the future as countries tighten up on sulfur emissions in a global effort to reduce acid rain.
Vol. 1016; pp. 30-35

Parrish, D.D.; Fahey, D.W.; Liu, S.C.; Fehsenfeld, F.C.; Hahn, C.H.; Williams, E.J.; Bollinger, M.J.; Huebler, G.; Buhr, M.P.; Murphy, P.C.; Trainer, M.; Hsie, E.Y.
Systematic variations in the concentration of NO_x ($NO + NO_2$) at Niwot Ridge, Colorado. Journal of Geophysical Research; Feb 20 1990. Description: Measurements of the concentrations of NO and NO_2 were made in the rural troposphere during a year's period in 1980-1981, during the summers of 1983, 1984, and 1987, and during the fall of 1984. The field site was located near Niwot Ridge, Colorado, at an elevation of 3 km. NO was measured with a chemiluminescence instrument, and NO_2 was photolyzed to NO and measured by the same instrument. The performance of this instrument is discussed in detail.
Vol. 952; pp. 1817-1836

Parrish, Donna.L.; Behnke, Robert.J.; Gephard, Stephen.R.; McCormick, Stephen.D.; Reeves, Gordon.H.
Why aren't there more Atlantic salmon (*Salmo salar*)?. Canadian Journal of Fisheries and Aquatic Sciences; Jan 01 1999. Description: Numbers of wild anadromous Atlantic salmon (*Salmo salar*) have declined demonstrably throughout their native range. The current status of runs on rivers historically supporting salmon indicate widespread declines and extirpations in Europe and North America primarily in southern portions of the range. Many of these declines or extirpations can be attributed to the construction of mainstem dams, pollution (including acid rain), and total dewatering of streams.
Vol. 56; no. 1; pp. 281-287

Patel, A; Chen, Y.P; Mishra, N.C.
Strategy for large scale solubilization of coal-characterization of neurospora protein and gene. Preprints of Papers, American Chemical Society, Division of Fuel Chemistry; Jan 31 1995. Description: Coal represents an important source of energy. Its utilization for generating energy has been offset by environmental problem that it creates by the release of SO_x and NO_x, which are major causes of acid rain and deforestation. Some of these problems can be tackled by the use of industrial scrubbers. However, a biotechnological approach to these problems may prove more efficient and environment friendly. We have employed certain genetically characterized fungi for the biosolubilization of coal in order to yield chemicals that can be converted into utilizable energy and can be rendered free of SO_x and NO_x at the source.
Vol. 40; no. 3; pp. 547-549

Patrizia, C.A.; Colbourn, E.L.; Wortham, G.L.
Regulatory conflicts facing electric utilities under the Clean Air Act amendments of 1990. Electricity Journal; Mar 31 1992. Description: Passage of the 1990 Clean Air Act amendments (CAAA, or amendments) will significantly affect the electric generation sector for decades to come. The Act is much broader than its acid rain and allowance trading provisions. The authors believe that in considering its effects on the utility industry, one must recognize the interrelated effects of several programs on the industry. One must also recognize that the implementation of the CAAA will be far more complex than a simple analysis (involving some combination of state, federal and utility interests) would indicate. State/federal and multistate issues are raised by the amended Clean Air Act as it effects electric utilities. Vol. 52; pp. 28-36

Patt, A.
Separating analysis from politics: Acid rain in Europe. Policy Studies Review; Dec 31 1999. Description: Over the last twenty years, policy-makers in Europe have attempted to solve the problem of acid rain using detailed analysis grounded in natural

science and economics. The results are impressive, as Europeans have successfully implemented a number of international agreements to reduce pollution emissions, agreements that in theory achieve the greatest environmental benefit at the lowest aggregate cost across Europe. This article examines the analysis on which these policies were based. Vol. 16; no. 3-4

Penner, J.E.; Atherton, C.S.; Dignon, J.; Ghan, S.J.; Walton, J.J.; Hameed, S.
Tropospheric nitrogen: A three-dimensional study of sources, distributions, and deposition. Journal of Geophysical Research; Jan 20 1991. Description: The authors simulate the global cycle of reactive nitrogen in a three-dimensional model of chemistry, transport, and deposition. The model is based on the Lagrangian tracer model and uses winds and precipitation fields calculated by the Livermore version of the NCAR Community Climate Model. The model includes the basic chemical reactions of NO, NO_2, and HNO_3. For this study, they use prescribed OH and O_3 concentrations and calculate the concentrations of NO, NO_2, and HNO_3 for a perpetual January and a perpetual July. The sources of reactive nitrogen due to fossil-fuel combustion (22 Mt N/yr), lightning discharges (3 Mt N/yr), soil microbial activity (10 Mt N/yr), biomass burning (6 Mt N/yr), and the oxidation of N_2O in the stratosphere (1 Mt N/yr) are included. In general, they find reasonable agreement between model predictions and measurements except for concentrations of HNO_3 in the remote Pacific.
Vol. 961; pp. 959-990

Peterman, R.M.
The importance of reporting statistical power: The forest decline and acidic deposition example. Ecology; Oct 31 1990 Description: The author uses papers which reported little evidence of the effect of acid deposition on forest ecosystems to point out the problems of statistical reporting practices in ecology. He suggests that often we are not given the information necessary to judge the strength of the evidence in these reports; i.e., their data analyses or experiments may have had power too low to warrant being used as evidence. Low power could result from sample sizes too small or data sets too variable to have a high chance of finding a statistically significant effect of acid deposition. Risk assessments relevant to natural resource management should be based on concepts of probability of type II error, power and detectable effect size.
Vol. 715; pp. 2024-2027

Peterson, C.E. Jr.; Mickler, R.A.
Considerations for evaluating controlled exposure studies of tree seedlings Journal of Environmental Quality; Mar 31 1994. Description: Tree seedling exposure studies, covering a wide range of experimental conditions in pollutant treatments, species, facilities, and exposure regimes, have been conducted during the past several years to determine acute effects and relative sensitivity of tree species in response to simulated acid precipitation and gaseous pollutants. Because of the difficulties inherent in conducting controlled exposures with mature trees (e.g., size, variability among experimental units, costs), seedling exposure studies have been initiated as the quickest way to address these issues.
Vol. 23; no. 2; pp. 257-267

Phase I of acid rain program is a success
Air Pollution Consultant; Jan 31 1997 Description: As of January 1, 1995, Phase I of the acid rain program required the 110 largest, high-emitting utility plants (located in 21 states) to comply with intermediate sulfur dioxide (SO_2) emission limitation requirements. These controls are designed to reduce SO_2 emissions from units to less than or equal to 2.50 pounds per million Btu (lb/MMBtu). According to a report recently released by EPA, all of the facilities affected by phase I successfully met their compliance obligations. The results of the Phase I performance evaluation were published in a July 1996 report entitled '1995 Compliance Results - Acid Rain Program' presented here in brief.
Vol. 7; no. 1; pp. 1.8-1.9

Phillips, K.
>   Where have all the frogs and toads gone
>   Bioscience; Jun 30 1990. Description: At a
>   recent workshop in California, scientists
>   discussed the decline of amphibian
>   populations and suggested that these
>   animals may be biological indicators of
>   advanced degradation of the environment.
>   One study described the effects of
>   snowmelt contamination with acid
>   deposition from smog and smelters on the
>   breeding ponds of salamanders. Other
>   possible reasons for decline include: heavy
>   metals and pesticides; global climate
>   changes; imbalances in mammal
>   populations that prey on amphibians;
>   predation by fishes stocked in lakes by
>   wildlife managers; and human predation.
>   Vol. 406; pp. 422-424

Phillips, R.A.; Stewart, K.M.
>   Longitudinal and seasonal water chemistry
>   variations in a northern Appalachian
>   stream. Water Resources Bulletin; Jun 30
>   1990. Description: Quaker Run, a fourth
>   order stream located in southwestern New
>   York State, exhibits a highly unusual
>   chemical gradient along its upper reaches.
>   Weekly water samples showed an increase
>   in the mean annual pH from 5.07 to 7.01
>   along a stretch of only 2.2 km. Mean
>   alkalinity, calcium, magnesium, sodium,
>   potassium, nitrate-nitrite-nitrogen, silica,
>   and conductivity also increased
>   appreciably over this distance. The study
>   area receives some of the most highly
>   acidic atmospheric deposition in the US.
>   Minimal buffering of these acidic inputs in
>   the extreme upper watershed, and an
>   abrupt downstream increase in buffering
>   associated with changes in soil type,
>   apparently produce the observed
>   streamwater chemistry gradient. In
>   contrast, a comparison between 11
>   midstream, downstream, and tributary sites
>   showed relatively little variation in
>   streamwater chemistry. In addition to the
>   pronounced longitudinal chemistry
>   changes along the upper portion of the
>   stream, pronounced temporal chemistry
>   variations were also observed at all
>   sampling sites. High flow during snowmelt
>   and heavy rains produced more dilute,
>   acidic conditions, while streamwater pH
>   and dissolved base cations were generally
>   highest during low flow.
>   Vol. 263; pp. 489-498

Phillips, V.D.
>   Living in a terrarium
>   Environmental Science and Technology;
>   Apr 30 1991. Description: This paper
>   offers suggestions for basic research that
>   needs to be completed to fill critical gaps
>   in scientific knowledge of global warming
>   and presents some basic concepts of the
>   relationship between green plants and
>   climate control, using oceanic
>   phytoplankton, tropical forests, and boreal
>   forests as examples. Also discussed is the
>   potential role of green plants in the
>   remediation of global warming through (1)
>   policy changes addressing deforestation,
>   acid rain, sustainable farming practices,
>   and reforestation, and (2) technical fixes
>   involving plant migrations, plant breeding,
>   and fertilizing the open oceans. Although
>   the measures presented here can provide
>   temporary assistance, a remedial course in
>   human ecology may be the ultimate
>   solution to our survival and quality of life.
>   Vol. 254; pp. 574-578

Pier, P.A.; Thornton, F.C.; Neufeld, H.; Seiler, J.R.; Hutcherson, J.D.
>   Cloudwater and O_3 effects on red spruce at
>   Whitetop Mt., VA: Physiological response
>   Bulletin of the Ecological Society of
>   America; Jun 30 1994
>   Description: Results of studies on red
>   spruce (Picea rubens Sarg.) at Whitetop
>   Mountain (elevation 1689 m) were
>   assessed to evaluate whether acidic
>   cloudwater deposition and O_3 contribute to
>   reported high elevation red spruce
>   ecosystem decline. Studies were conducted
>   using seedling exclusion chambers, mature
>   tree branch exclusion chambers, and field
>   experiments with seedlings, saplings, and
>   mature trees. Ozone had minimal effects
>   on the measured parameters.
>   Photosynthetic response to cloudwater
>   varied, dependent on tree age.
>   Vol. 752; pp. 181

Pleim, J.E.; Chang, J.S.; Zhang, Kesu
A nested grid mesoscale atmospheric chemistry model. Journal of Geophysical Research; Feb 20 1991. Description: A nested grid version of the Regional Acid Deposition Model (RADM) has been developed. The horizontal grid interval size of the nested model is 3 times smaller than that of RADM (80/3 km = 26.7 km). Therefore the nested model is better able to simulate mesoscale atmospheric processes while maintaining consistency with larger-scale features. Vol. 962; pp. 3065-3084

Plummer, P.L.M.
A theoretical investigation of HSO/HOS and their positive ions. Journal of Chemical Physics; Jun 01 1990 Description: The formation and eventual fate of sulfur-containing aerosols play a central role in global pollution. An understanding of the oxidation paths for sulfur species and of the formation and stability of radical and ionic intermediates is required for optimum control of acid deposition. To gain insight into these processes *ab initio* calculations were performed for ground and first excited electronic states of the isomers HSO/HOS and for the ground and first two excited states for their positive ions, HSO^+/HOS^+. A variety of basis sets were used for calculations at the self-consistent field (SCF) level. These calculations included full optimization of the geometry and examination of the potential surface for transition states. Vol. 9211; pp. 6627-6634

Polluting a microbial methane sink
Bioscience; Mar 31 1990
Description: Excess nitrogen, whether from fertilization or from acid rain, seems to reduce the amount of methane that soil organisms can remove from the atmosphere. Methane, an important greenhouse gas, contributes to global warming by acting as an atmospheric blanket. The gas has been increasing approximately 1% a year for the past decade, due either to increases in global sources or decrease in biological sinks. The largest such sinks are the microorganisms in aerobic soils. Recent research, has shown that added nitrogen significantly decreases the rates at which temperate forest soils can take up methane. Laboratory studies with soil microorganisms support the field observations, suggesting that high nitrogen suppresses methane uptake. The researchers say further measurements in agroecosystems, pastures, and other high-nitrogen systems are needed to clarify the nitrogen-methane interaction before extrapolation to a global basis. Vol. 403; pp. 230

Ponge, Jean-François.; Arpin, Pierre.; Sondag, Francis.; Delecour, Ferdinand.
Soil fauna and site assessment in beech stands of the Belgian Ardennes Canadian Journal of Forest Research; Dec 01 1997. Description: Soil fauna (macrofauna and mesofauna) were sampled in 13 beech forest stands of the Ardenne mountains (Belgium) covering a wide range of acidic humus forms. The composition of soil fauna was well correlated not only with humus form, but also with elevation, phytosociological type, tree growth, mineral content of leaf litter, and a few soil parameters such as pH and C/N ratio. The nature of the mechanisms that can explain these relationships is discussed in light of existing knowledge. Vol. 27; no. 12; pp. 2053-2064

Pool, R.
User-friendly chemistry takes center stage at ACS meeting. Science (Washington, D.C.); Sep 11 1992. Description: These days it seems that what chemistry needs more than anything else is a good p.r. agent. If you ask John or Joan Q. Public about the accomplishments of the chemical industry, chances are they'll mention Love Canal, CFCs destroying the ozone layer, or carcinogens in food. However, if the national meeting of the American Chemical Society in Washington, D.C., 2 weeks ago is any indication, chemists are working hard to fix the image problem. Vol. 2575076; pp. 1479-1480

Popova, T.V; Vishnyakova, T.P; Yurechko, V.V; Frolov, V.I
Oxidizability and stabilization of ecologically clean diesel fuel. Chemistry and Technology of Fuels and Oils; Nov 30 1995. Description: The dieselization of the automotive fleet of this country and the growing demand for diesel fuel have put two problems into sharp focus: how to meet the demand for fuel, and how to satisfy stringent requirements on the exhaust smoke level and toxicity. The combustion of sulfur compounds present in diesel fuel forms sulfur oxides that are responsible for metal corrosion, attack on roads and buildings, acid rain, and other detrimental phenomena. In order to improve the ecological situation, particularly in large cities suffering from constantly increasing emissions from automotive transport operations (which account for up to 80% of the total atmospheric pollution), a diesel fuel with a lower sulfur content has been developed, the so-called ecologically clean diesel fuel DLECh. Vol. 31; no. 3-4; pp. 116-120

Popp, C.J.; Lynch, T.R.; Jacobi, G.Z.
Acidification potential in high mountain lakes in northern New Mexico American Chemical Society, Division of Environmental Chemistry, Preprints; Jan 31 1990. Description: Much of the concern related to the environmental impact of acidic precipitation has focused on areas east of the Mississippi River, particularly in the upper Midwest, New England, and southeastern Canada. Evidence that acidic precipitation is a potential problem in various parts of the West has been mounting. In recent years, acidic precipitation events have been detected with regularity in the Sierras, the Cascades, central Colorado, and in northern and central New Mexico. The specific objectives of the research were to: (1) document the present pH of high mountain wilderness lakes, (2) determine the vulnerability of these lakes to future acidification, (3) determine using chemical and biological procedures the loss of buffering capacity, (4) establish via sediment cores a chronological record of trace metal deposition from atmospheric inputs, and (5) study the monthly variation of pH, alkalinity and other chemical parameters as a function of precipitation inputs and biological activity. Vol. 301; pp. 184-185

Powell, B.E.
Seven summers of energy education courses. Georgia Journal of Science; Jan 31 1995. Description: Beginning in 1988 and continuing until 1994, seven grants were obtained from the Office of Energy Resources of the State of Georgia to offer energy education courses to pre-college teachers. The grants totalled about $169,000, and 220 teachers participated in the courses. The courses covered the historical usage and sources of energy, current energy sources and supplies of energy, environmental concerns (including acid rain, the greenhouse effect, and the depletion of atmospheric ozone), energy conservation, alternate energy sources, research to develop new energy sources, classroom demonstrations, and applications. Vol. 53; no. 1; pp. 48

Power, M.
Decision-making for the restoration of Atlantic biology and economics Canadian Journal of Fisheries and Aquatic Sciences; Jan 01 1998. Description: Acidification of Atlantic salmon (*Salmo salar*) rivers represents a major threat to salmon production in much of Nova Scotia, Canada. Efforts at understanding the efficacy of proposed remedial strategies have concentrated on estimating the biological parameters of the acidification issue. However, the dominance of societal values in the allocation of resources to fisheries management problems demands alternative strategies for the remediation of acidity in salmon rivers be developed that account for both the biological and cost constraints on remedial strategy selection. Vol. 55; no. 1; pp. 143-149

Prenzel, J.
Sulfate sorption in soils under acid deposition: Comparison of two modeling

approaches. Journal of Environmental Quality; Jan 31 1994. Description: Soil acidification under the impact of acid deposition is often modeled by adsorption isotherms. The assumption of solubility equilibria for basic Al sulfates is an alternative modeling approach. Similarities and differences between both concepts are discussed. A simple solubility equilibrium model is defined and used to demonstrate that the second model can explain results that are well described by typical adsorption isotherms. The same model can also explain a pH dependency of adsorption isotherms that has been widely reported. In the application to assumed soil acidification scenarios, predictions are presented that could not be described by an adsorption isotherm approach.
Vol. 23; no. 1; pp. 188-194

Prenzel, J; Meiwes, K.J.
Sulfate sorption in soils under acid deposition: Modeling field data from forest liming. Journal of Environmental Quality; Nov 30 1994. Description: Sulfate sorption can modify the reaction of a soil to acid depositions. Different models of this process exist. In a preceeding paper the adsorption isotherm and the solubility product modeling approach have been compared and a solubility product model has been introduced. The latter model is tested using a forest liming experiment. Concentration data form canopy drip and seepage covering a 20-yr observation period are used as 3-yr moving averages.
Vol. 23; no. 6; pp. 1212-1217

Quillian, A.M.; Lundgren, D.A.
Field measurements of dry deposition compounds using the transition flow reactor. Journal of the Air and Waste Management Association; Jan 31 1992. Description: The transition flow reactor (TFR) was used to measure ambient concentrations of particulate nitrate (NO_3^-), particulate sulfate ($SO_4^=$), nitric acid (HNO_3), and sulfur dioxide (SO_2) around the Gainesville, Florida area. The TFR measured concentrations were then compared to measurements made by different filter packs. Vol. 421; pp. 36-39

Rabl, V.A.
Beneficial electrification: Environmental advantages of new electricity uses
IEEE Power Engineering Review (Institute of Electrical and Electronics Engineers); Nov 30 1993. Description: During the past decade, there has been a rapid rise in public awareness and concern with the need for environmental protection. Widespread regulatory initiatives have been put in place to conserve energy resources and take action to stem a variety of environmental issues such as urban air pollution, acid rain, greenhouse gases, and depletion of the upper atmosphere's protective ozone layer. Through its generation and use, electric power can be one of the many sources of emissions that contribute to some of these problems.
Vol. 1311; pp. 3-7

Rafizadeh, H.A.
Multi-disciplinary management needs of power systems: A new challenge
Proceedings of the American Power Conference; Jan 31 1992
Description: Drastic shifts are taking place in the utility risk portfolio. The currently localized technical, operational, and regulatory risks are changing to long-term risks that are global in character. Specifically, these risks are: global warming; acid rain; electromagnetic fields; and open market competition. This paper reports that they are all related to the production and use of electricity and in aggregate require a new multi-disciplinary style of management at the utility companies. Vol. 541; pp. 377-382

Rahman, S; Castro, A. de
Environmental impacts of electricity generation: A global perspective
IEEE Transactions on Energy Conversion; Jun 30 1995
Description: Efforts are now underway in the industrialized countries to significantly reduce the emission of greenhouse and acid rain gases from power plants. However, indications are that rapid growth in the developing countries is quickly replacing what is being eliminated. This paper provides data and projections on these

emissions from electricity generation in selected countries around the world, makes some comparisons and suggest possible options. Vol. 10; no. 2; pp. 307-314

Rajaretnam, G; Blasio, C; Lovins, K; Spitz, H.B.
Effect of leachability on environmental risk assessment for naturally occurring radioactive materials (NORM) in petroleum oil fields. Health Physics; Jun 30 1996. Description: Elevated concentrations of NORM often occur in petroleum oil fields. The NORM generated by oil field operations comes from ^{238}U and ^{232}Th contained in geologic materials. The predominant NORM radionuclide brought to the surface by produced water is radium, which co-precipitates with barium in the form of complex compounds of sulfates, carbonates, and silicates in sludge and scale. These NORM deposits are highly stable and very insoluble under ambient conditions at the surface.
Vol. 70; no. Suppl.6; pp. 36a

Ramsay, S. L.; Houston, D. C.
Do acid rain and calcium supply limit eggshell formation for blue tits (Parus caeruleus) in the U.K.?
Journal of Zoology; Jan 01 1999
Vol. 247; no. 1; pp. 121-125

Rees, T.H.; Schnoor, J.L.
Long-term simulation of decreased acid loading on forested watershed
Journal of Environmental Engineering (New York); Jan 31 1994
Description: The response of a northeastern US hardwood forest to wet and dry deposition of hydrogen ion and sulfate deposition was performed using the hydrobiogeochemical, lumped-parameter Terrestrial Aquatic Model for Ecosystems (TAME). Results of model calibration and output stream chemistry show that ion-exchange and mineral dissolution processes in the B and C horizons buffer the acidic, ambient atmospheric deposition [864 eq-SO_4^{-2}/(ha/yr) and 718 eq-H^+/(ha/yr)] in this hydrologically tight and flashy watershed. Vol. 1202; pp. 291-312

Regina, K.; Nykänen, H.; Maljanen, M.; Silvola, J.; Martikainen, P.J.
Emissions of N_2O and NO and net nitrogen nitrogen compounds. Canadian Journal of Forest Research; Jan 01 1998. Description: Fluxes of nitrous oxide (N_2O) and nitric oxide (NO) were measured on a drained and forested peatland in 1992-1995. Net mineralization and nitrification were studied in situ in 1993-1994. Nitrogen additions in 1992 as KNO_3, NH_4Cl, or urea (100 kg N ·ha^{-1}) were used to study the fate and transformations of N in peat. The mean N_2O emissions during the growth season in 1993 were 1.9, 2.6, 3.3, and 3.5 mg N ·m^{-2} ·day^{-1} in the control soil, KNO_3, NH_4Cl, and urea-treated soils, respectively.
Vol. 28; no. 1; pp. 132-140

Renner, R
'Scientific uncertainty' scuttles new acid rain standard. Environmental Science and Technology; Oct 31 1995. Description: An EPA report to Congress due this month will report on the controversial question of whether the Clean Air Act Amendments of 1990 (CAAA) adequately protect sensitive areas of the United States from acid rain, and recommends against establishing a new 'acid deposition standard' to protect sensitive areas of the United States from acid rain. Rebecca Renner reports on the scientific issues underlying that decision and the efforts of one state to overturn it. The report to Congress, required by the CAAA, asked the Agency to report on the feasibility of setting an acid deposition standard to protect sensitive areas.
Vol. 29; no. 10; pp. 464A-466A

Report of the Committee on the environment
Energy Law Journal; Jan 31 1994
Description: This report of the Committee on the Environment focuses on a variety of environmental matters of interest to energy practitioners. First, it notes various important regulatory and litigation activities concerning the Acid Rain Program under the Clean Air Act Amendments of 1990. A discussion on pollution prevention and various industry and EPA initiative follows. Third is Global Climate Change, including a description of

the Clinton Administration's Climate Change Action Plan. Finally, there is an update on environmental externalities and a report on the EPA's determination under RCRA concerning fossil fuel combustion wastes. Vol. 151; pp. 175-191

Report of the Committee on the Environment
Energy Law Journal; Jan 31 1992 Description: In 1991 the Environmental Protection Agency (EPA) labored to develop an enormous body of rules to implement the Clean Air Act Amendments of 1990. While the EPA made progress, developing draft rules for some programs and publishing proposed rules for others, it did not issue final rules for the ozone nonattainment, toxics, and acid rain programs. Discussed in this report are the Acid Rain Permit Rule, the proposed Allowance System rule, the proposed Continuous Emissions Monitoring Rule, Wisconsin Electric Power Co. v. Reilly (WEPCO) rulemaking, the Clean Air Act Amendments of 1990, the National Energy Security Act of 1992, RCRA reauthorization and oil and gas explorations and production wastes, and congressional action.
Vol. 131; pp. 141-157

Reynolds, W.E.
Care and feeding of continuous emissions monitoring systems. Environmental Protection; Jul 31 1995. Description: If you thought you could kick back and relax now that you've successfully certified your new Continuous Emissions Monitoring Systems (CEMS), think again. The CEMS you installed to comply with EPA's Acid Rain Reduction Program does more than document compliance with emissions limitations; it also meters the flow of money out of the corporate bank account. Indeed, its accuracy has a significant effect on your bottom line. Why? Because under the emissions trading program created by the 1990 Clean Air Act Amendments, SO_2 emissions translate into dollars and cents on the commodities market.
Vol. 6; no. 7; pp. 19-21

Rhorer, J.R. Jr; Warren, P.R.
Force majeure implications of acid rain legislation: The litigation battle of the 1990s. Journal of Energy, Natural Resources amp Environmental Law; Jan 31 1992. Description: During the 1980s, utilities turned to the courts to avoid the high coal prices they agreed to pay under long-term, "base price plus escalator" contracts entered into during the energy crisis of the 1970s, or shortly thereafter. When spot market coal prices and energy demand fell in the early 1980s, but contract prices continued to escalate, utilities carefully evaluated their contracts to determine if there might be some excuse from performance that a regulatory body, in hindsight, could use to deny a rate increase. In the words of one court, a utility "apparently searched the law books to unearth every conceivable cause of action." Vol. 8; no. 1; pp. 23

Richardson, G.M.; Egyed, M.; Currie, D.J.
Does acid rain increase human exposure to mercury A review and analysis of recent literature. Environmental Toxicology and Chemistry; May 31 1995. Description: The literature suggests that acid deposition may lead to increased mercury (Hg) contamination of fish. Employing published empirical relationships, the authors have estimated the change in associated Hg contamination with an increase in sulfate deposition from 0.25 to 1.25 g sulfur/m^2/year. In seepage lakes, one can predict that Hg in walleye from these lakes, and subsequent human exposure due to consumption of these fish, would be elevated at the higher rate of sulfate deposition. However, for drainage lakes, increasing acidic deposition was predicted to reduce Hg accumulation in lake trout and northern pike. Vol. 145; pp. 809-813

Rickett, B.I; Payer, J.H.
Composition of copper tarnish products formed in moist air with trace levels of pollutant gas: Hydrogen sulfide and sulfur dioxide/hydrogen sulfide. Journal of the Electrochemical Society; Nov 30 1995. Description: The objective of the work here was to characterize the tarnish

products formed on copper during the early stages of exposure to moist air with trace levels of pollutant gas, in particular hydrogen sulfide and sulfur dioxide/hydrogen sulfide. As determined by X-ray photoelectron spectroscopy and coulometric reduction (chronopotentiometry), the exposure of copper to 50 ppb hydrogen sulfide at 23 C and 70% relative humidity was found to produce a tarnish layer consisting of Cu_2O and Cu_2S. Addition of 75 ppb sulfur dioxide to a 50 ppb hydrogen sulfide, moist air environment at 23 C and 70% relative humidity encouraged the growth of a thicker tarnish product layer, with proportionally more Cu_2O in the Cu_2O/Cu_2S structure.
Vol. 142; no. 11; pp. 3723-3728

Rinallo, C.
Effects of acidity of simulated rain on the fruiting of Summerred' apple trees Journal of Environmental Quality; Jan 31 1992. Description: The effects of rain acidity on field-grown Summered apple trees (Malus domestica Borkh) under natural conditions were investigated. One group of four trees was exposed to ambient rainfall. Four other groups were covered with rainshields and received water, pH 5.6, 4, and 3, respectively, as simulated rain. Simulated acid rain, particularly at pH 3, adversely affected fruit production in terms of individual fruit weight, fruit set, fruit appearance (necrosis and russetting of the peel) & dry weight. Vol. 211; pp. 61-68

Rittenhouse, R.C.
Action builds on 1990 Clean Air Act compliance. Power Engineering; May 31 1992. Description: This paper reports on some of the many Clean Air Act, (CAA) deadlines. Phase I of SO_2 emissions reductions begins January 1, 1995 and affects 261 units in 110 coal-burning power plants in 21 eastern and midwest states. Phase II begins January 1, 2000 and imposes tighter emissions limits on these plants and sets restrictions on smaller, cleaner plants fueled by coal, oil and gas. This affects 2500 units at 1000 power plants. Continuous emission monitors (CEMs) must be in place on Phase I units by November 15, 1993 and on Phase II units by January 1, 1995. In addition, the legislation calls for a 2-million-ton cut in NO_x emissions by 2000. This is to come from demands under the acid rain requirements (Title IV) of the CAA and is in two parts. The EPA is to set emission limits for tangential-fired and dry-bottom boilers by mid-1992 and all other boilers by 1997. CEMs also are required for NO_x emissions. Vol. 965; pp. 21-28

Rodgers, L.M.
NARUC winter meetings address key issues for utility industry. Public Utilities Fortnightly; Apr 12 1990. Description: This article reports on the National Association of Regulatory Utility Commissioners (NARUC) meeting of February, 1990. The topics covered by the report include acid rain (including emissions trading), electric and magnetic fields (as air pollution), energy conservation (including a call for increased funding of research, development and commercialization of energy conservation and renewable energy technologies), natural gas, and telecommunications.
Vol. 1258; pp. 26-28

Rondon, A.; Johansson, C.; Granat, L.
Dry deposition of nitrogen dioxide and ozone to coniferous forests. Journal of Geophysical Research; Mar 20 1993. Description: The exchange of NO_2, NO, and O_3 between the atmosphere and coniferous forests has been studied by using a dynamic flow-through chamber technique. The measurements were performed during summer at two coniferous forest sites in Sweden, Jaedraas (Scots pine) and Simlaangsdalen (Scots pine and Norway spruce). In Simlaangsdalen, the flux of NO_2 was found to be quantitatively determined by the stomatal openings.
Vol. 983; pp. 5159-5172

Rooth, R.A.; Verhage, A.J.L.; Wouters, L.W.
Photoacoustic measurement of ammonia in the atmosphere: influence of water vapor and carbon dioxide. Applied Optics; Sep

01 1990. Description: The photoacoustic determination of the ammonia concentration in atmospheric air by absorption of CO_2 laser radiation at 9.22 {mu}m is influenced by the presence of H_2O and CO_2. Kinetic cooling due to the coupling of excited CO_2 and N_2 levels causes important changes in phase and amplitude of the photoacoustic signal. Theoretical background is presented to deduce the correct NH_3 concentration from the signal. The experimental setup used to perform field measurements is described. Adhesion of NH_3 to the walls of the resonant photoacoustic cell was investigated. Temperature effects are treated. Field data of NH_3 and H_2O concentrations are presented. Key words: Photoacoustics, ammonia, kinetic cooling, trace gas measurements, ammonia adhesion, acoustic resonance, CO_2 laser radiation, water vapor absorption, carbon dioxide absorption.
Vol. 2925; pp. 3643-3653

Rossin, A.D.
Requirements for a viable nuclear future. Transactions of the American Nuclear Society; Jan 31 1991. Description: Different political perspectives have resulted in varied load projections of future electricity demand. However, most projections show that there will be a need for additional electrical base-load capacity by the end of the century to meet future demand. With emissions of greenhouse gases already a matter of international policy (and probably a future national policy issue in the United States) and with new acid rain laws in force, common sense calls for orders of new nuclear power plants as soon as possible.
Vol. 63; pp. 416-417

Rostam-Abadi, M.; Moran, D.L.
High-surface-area hydrated lime for SO2 control. Geological Society of America, Abstracts with Programs; Mar 31 1993 Description: Since 1986, the Illinois State Geological Survey (ISGS), has been developing a process to produce high-surface-area hydrated lime (HSAHL) with more activity for adsorbing SO2 than commercially available hydrated lime. HSAHL prepared by the ISGS method as considerably higher surface area and porosity, and smaller mean particle diameter and crystallite size than commercial hydrated lime. The process has been optimized in a batch, bench-scale reactor and has been scaled-up to a 20- - 100 lb/hr process optimization unit (POU).
Vol. 253; pp. 76

Rostorfer, C.R.
Keys to fuel supply success. Public Utilities Fortnightly; Jun 15 1991. Description: This article examines the changes to the fuel procurement process, some brought about by acid rain legislation, and provides a step-by-step guide to handling the changes. The topics include requirements planning, market research, developing a good procurement strategy, implementing the strategy, effective administration, and a checklist for contract review. Vol. 12712; pp. 20-22

Roy, K.A.
Nalco technology reduces nitrogen oxide emissions. Hazmat World; May 31 1991. Description: The 1990 CAA amendments sent many operators of stationary combustion sources searching for methods to reduce a variety of emissions, including nitrogen oxides (NO_x) - major components of acid rain. Their searches will reveal several strategies, including one - NO_xOUT - offered by Nalco Fuel Tech (Naperville, Ill.), and European-based Fuel Tech N.V. Nalco Fuel Tech was formed in February 1990 to market NO_xOUT, as well as specialty chemicals and services for air pollution control and other fuel-treatment-related needs. NO_xOUT is a selective non-catalytic reduction (SNCR) technology that combines process modifications and specialty chemicals to reduce NO_x. Specialty reducing chemicals are injected into flue gases, where they react with NO_x to form nitrogen, carbon dioxide and water.
Vol. 4; pp. 76-78

Roy, K.A.
New process removes sulfur from coal before burning. Hazmat World; Dec 31

1990. Description: This paper reports on technology to eliminate sulfur dioxide emissions generated by burning coal stepped to the front of the line. A desulfurization process removes sulfur from coal before it is burned, eliminating or minimizing sulfur dioxide emissions, an acid rain precursor. Historically, research has focused on post-combustion flue-gas desulfurization (FGD) techniques to reduce sulfur dioxide emissions.
Vol. 312; pp. 43-45

Rozanov, I.A; Kut'kov, V.S; Murashov, D.A.
Screening of sorbents for sensors to control sulfur dioxide content in air. Russian Journal of Coordination Chemistry; May 31 1995. Description: Features of the electronic structure of sulfur dioxide, hexaethoxycyclotriphosphazene $N_3P_3(OC_2H_5)_6$ molecules, and some sorbents examined earlier in piezochemical [1] sulfur dioxide sensors are considered in terms of the possible advantages of the reversible interaction of SO_2 with $N_3P_3(OC_2H_5)_6$. To compare the sensitivity of the materials to sulfur dioxide, the experimental sorption-desorption parameters for the reversible interaction of SO_2 with $N_3P_3(OC_2H_5)_6$ and a series of other sorbents applied in chromatographic columns as stationary phases are measured by means of gas chromatography. Being selected on the basis of sorption low energy and its sufficient reversibility, hexaethoxycyclotriphosphazene is advanced as a thin-film material for piezochemical sensors and a stationary phase in gas-liquid chromatography.
Vol. 21; no. 5; pp. 345-350

Rubiera, F; Arenillas, A; Fuente, E; Pis, J.J.; Marteinz, O; Moran, A.
Biodesulfurization of coals of different rank: Effect on combustion behavior. Environmental Science and Technology; Feb 01 1999. Description: The emission of sulfur oxides during the combustion of coal is one of the causes, among other air pollution problems, of acid rain. The contribution of coal as the mainstay of power production will be determined by whether its environmental performance is equal or superior to other supply options. In this context, desulfurization of coal before combustion by biological methods was studied. Vol. 33; no. 3; pp. 476-481

Rubin, E.S.
Benefit-cost implications of acid rain controls: An evaluation of the NAPAP integrated assessment. Journal of the Air and Waste Management Association; Aug 31 1991. Description: Concluding ten years of study, the US National Acid Precipitation Assessment Program (NAPAP) recently issued its integrated assessment report designed to provide guidance to policy makers on the sources and effects of acid deposition, and the costs and benefits of alternative control measures. This paper focuses on an evaluation of the benefit-cost implications of acid rain controls as revealed by two of the five major questions addressed in the NAPAP assessment framework. While the NAPAP effort made significant scientific contributions to the study of acid deposition, key gaps are found in the assessment of benefits and costs most relevant to policy decisions. Lessons learned from NAPAP may be helpful in avoiding similar problems assessing environmental issues such as global climate change. Vol. 418; pp. 914-921

Rubin, E.S.; Lave, L.B.; Morgan, M.G.
Keeping climate research relevant Issues in Science and Technology; Jan 31 1992. Description: Recent post-mortems of the National Acid Precipitation Assessment Program (NAPAP) confirmed what Congress and other key parties to the acid rain debate already knew: that the 10-year, half-billion-dollar interagency program to guide US policy on acid rain control proved largely irrelevant to the effort to forge the new Clean Air Act last fall. Although NAPAP won praise for its scientific accomplishments, the program failed in its primary mission - to provide policy- relevant information in a timely manner. Vol. 82; pp. 47-55

Ruch, R.B. Jr.; Howell, J.S. Jr.
Proactive industrial strategies for the Clean Air Act amendments of 1990
Journal of the Air and Waste Management Association; Aug 31 1991
Description: The Clean Air Act (CAA) Amendments of 1990 was signed into law by President Bush on November 15, 1990. These amendments potentially will have a major impact on virtually every industrial and many commercial facilities throughout the country. The regulations developed to implement this legislation will encompass new approaches to nonattainment, air toxics, accidental releases, acid rain, permits and enforcement. Because of the impact of this legislation the regulations will be implemented over a ten-year period. This paper is an overview of the amendments and recommended proactive strategies for industry.
Vol. 418; pp. 922-927

Russell, M.
NAPAP: A lesson in science, policy
Forum for Applied Research and Public Policy; Jan 31 1993. Description: Perplexing environmental questions, such as acid rain and global warming, cry out for policy solutions based upon solid scientific evidence. Scientists and politicians agree on this but have trouble finding an effective way to do it. Milton Russell of the University of Tennessee and Oak Ridge National Laboratory describes a major, but only partially successful, effort that he believes contains valuable lessons for scientists and policy makers in the future. Vol. 82; pp. 55-60

Ryan, B.D.; Nash, T.H. III
Lichen flora of the Eastern Brook Lakes watershed, Sierra Nevada Mountains, California. Bryologist; Jan 31 1991. Description: This report describes a quantitative study of lichens in the Easter Brook Lakes Watershed, in the Sierra Nevada Mountains (hereafter called Sierras) of California. The physical, chemical, and biological features of this watershed are currently being studied by several investigators in various disciplines, and the lichen study will contribute to overall knowledge of the site. Littler previous work has been done on the lichens of the Sierras, and the study reported here gives baseline data that will be useful to future studies of lichens in relation to acid rain or other pollution in the Sierras. . Vol. 942; pp. 181-195

Sadler, B.
Shared resources, common future: Sustainable management of Canada-United States border waters. Natural Resources Journal; Jan 31 1993. Description: A long tradition of transboundary resource management activities links the United States with Canada and with Mexico, especially with respect to shared waters. The institutions established for this purpose, notably the International Joint Commission (IJC) and the International Boundary and Water Commission (IBWC), have solid records of accomplishment. In recent years, however, the performance of both organizations has come under critical scrutiny. Vol. 332; pp. 375-396

Samson, P.J; Sillman, S.
A meteorology-based approach to detecting the relationship between changes in SO2 emission rates and precipitation concentrations of sulfate. Journal of Applied Meteorology; Sep 30 1994. Description: In this paper, the authors present an analysis of correlations between SO2 emissions and wet SO_4^{2-} concentrations over eastern North America that includes adjustments for the impact of meteorological variability. The approach uses multiple-regression models and readily available meteorological information to analyze precipitation chemistry data collected from 1979 to 1986 at six Utility Acid Precipitation Study Program sites. On an event-to-event basis, from 25% to 50% of the variation in concentrations, depending on site, was found to be related to meteorology. Precipitation amount, temperature, upwind emissions, and upwind mean lower-tropospheric relative humidity (indicator for upwind precipitation) were related to the natural log of SO_4^{2-} concentrations. Vol. 33; no. 9; pp. 1050-1066

Sandell, M.
> Putting NO$_x$ in a box. Pollution Engineering; Mar 31 1998. Description: Nitrogen oxide (NO$_x$) emissions are a major factor in national environmental problems, including acid rain and high ground-level ozone concentrations. As recently as 1996, more than 50 million Americans were living with unhealthy ozone levels. NO$_x$ also plays a role in elevated fine particulate levels, a pollutant which the US Environmental Protection Agency (EPA) intends to lower by revising the National Ambient Air Quality Standards (NAAQS). Numerous technologies are used to control NO$_x$. Vol. 30; no. 3; pp. 56-58

Sarosdy, R.L.
> May acid rain legislation excuse performance obligations under coal contracts. Energy Law Journal; Jan 31 1993. Description: The Clean Air Act Amendments of 1990 (CAAA) require 111 electric utility plants to reduce significantly their emissions of sulfur dioxide and nitrogen oxides by January 1, 1995. Further reductions of these emissions by all electric utilities will be required prior to January 1, 2000. Many electric utilities find themselves unable to utilize coal purchased under long-term contracts before the enactment of the CAAA unless they make significant and costly modifications to their generating facilities and/or purchase sulfur dioxide allowances. Vol. 142; pp. 303-333

Sase, Hiroyuki.; Takamatsu, Takejiro.; Yoshida, Tomio.
> Variation in amount and elemental composition of leaves associated with natural environmental factors. Canadian Journal of Forest Research; Jan 01 1998. Description: Leaf samples of *Cryptomeria japonica* D. Don (and some other conifers) taken from various locations in Japan were analyzed for differences in the amount and elemental composition of their epicuticular wax. In *C. japonica* the amount of wax per unit leaf mass was lower, and the C content of the wax relatively higher, than those of other species. The properties of the wax (amount, C and O contents) varied according to natural environmental factors such as altitude and exposure to volcanic acidic gases such as H$_2$S, as well as branch height and leaf age within the tree. Vol. 28; no. 1; pp. 87-97

Sase, Hiroyuki.; Takamatsu, Takejiro.; Yoshida, Tomio.; Inubushi, Kazuyuki.
> Changes in properties of epicuticular wax and the affected by anthropogenic environmental factors. Canadian Journal of Forest Research; Apr 01 1998. Description: The leaves of *Cryptomeria japonica* D. Don collected near an electrochemical plant (on Yakushima Island) had more wax (approximately 10% higher in 1-year leaves) and less chlorophyll (approximately 50 and 30% lower in 0- and 1-year leaves, respectively) than those from a reference area, although the trees showed no symptoms of decline. In the Kanto Plain around Tokyo (Saitama and Ibaraki), where *C. japonica* is declining (dieback and (or) defoliation), the amount of epicuticular wax in current-year leaves and the leaf chlorophyll content were almost equivalent to those of healthy plants in mountainous areas, but the wax eroded more rapidly (approximately 1.5 times faster). Vol. 28; no. 4; pp. 546-556

Sasek, T.W.; Richardson, C.J.; Fendick, E.A.; Bevington, S.R.; Kress, L.W.
> Carryover effects of acid rain and ozone on the physiology of multiple flushes of loblolly pine seedlings. Forest Science; Sep 30 1991. Description: The effects of acid rain and ozone exposure on loblolly pine (Pinus taeda L.) seedlings in the Piedmont of North Carolina were assessed over two exposure seasons (1987-1988). Direct effects and carryover effects of long-term exposure on the photosynthetic potential and photopigment concentrations of different needle age-classes were studied. Three half-sib families were grown in open-top field chambers and exposed two acid rain treatments and five ozone exposures delivered in proportion to ambient concentrations in a complete factorial design. Vol. 374; pp. 1078-1098

Saunders, G.L.; Laznow, J.
   The Clean Air Act Amendments of 1990 and industry: Title I non-attainment areas Hazmat World; Oct 31 1991. Description: The signing into law of the CAA Amendments of 1990 will bring sweeping changes affecting significantly the way industry is regulated. This paper reports that the Amendments address a wide range of issues, including non- attainment areas, toxic air pollutants, acid rain, operating permits and fees, and regulatory enforcement. Regulations to be promulgated under the Amendments will stand in stark contrast to those promulgated after CAA was last amended in 1977. Many of the issues addressed by the 1990 Amendments have accumulated for years, waiting for legislation consideration.
   Vol. 410; pp. 46-47

Savastano, C.A.
   The solvent extraction approach to petroleum demetallation. Fuel Science and Technology International; Jan 31 1991. Description: Metals are present in petroleum depending on age and conditions of diagenesis, and concentrate during refining in heavy fractions and residua. Nickel and vanadium show particularly deleterious effects on catalysts, such as poisoning, excessive gas and coke formation. Besides, removing vanadium from fuel oils reduces the environmental impact of acid rain due to the oxidation of sulphur dioxide in the atmosphere. In this paper, research and industrial practice literature concerning the removal of nickel and vanadium from petroleum and its fractions by solvent extraction and related techniques is reviewed.
   Vol. 97; pp. 855-872

Saylor, R.D.
   Atmospheric chemistry research Energeia (Lexington, Kentucky); Jan 31 1990. Description: Global environmental changes are occurring all around us, and the energy industry is a major player in the changes that are taking place. Wise energy policy can only be generated from a position of informed enlightenment and understanding about the environmental consequences of energy production and utilization. The atmospheric chemistry research being conducted at the University of Kentucky's Center for Applied Energy Research is geared toward providing the knowledge necessary to allow industrial and legislative officials to make responsible energy decisions in the 1990's and beyond. Three programs are described: the Kentucky Acid Deposition Program Precipitation chemistry network; modeling of regional and urban photochemistry and acid deposition; and modeling of global tropospheric chemistry. Vol. 11; pp. 1-2

Schaefer, D.A.; Driscoll, C.T. Jr.; R., Van Dreason,; Yatsko, C.P.
   The episodic acidification of Adirondack lakes during snowmelt. Water Resources Research; Jul 31 1990. Description: Maximum values of acid neutralizing capacity (ANC) in Adirondack, New York lake outlets generally occur during summer and autumn. During spring snowmelt, transport of acidic water through acid-sensitive watersheds causes depression of upper lake water ANC. In some systems lake outlet ANC reaches negative values. The authors examined outlet water chemistry from 11 Adirondack lakes during 1986 and 1987 snowmelts. In these lakes, SO_4^{2-} concentrations were diluted during snowmelt and did not depress ANC. For lakes with high baseline ANC values, springtime ANC depressions were primarily accompanied by basic cation dilution. For lakes with low baseline ANC, NO_3^- increases dominated ANC depressions. Lakes with intermediate baseline ANC were affected by both processes and exhibited larger ANC depressions. Ammonium dilution only affected wetland systems. A model predicting a linear relationship between outlet water ANC minima and autumn ANC was inappropriate. To assess watershed response to episodic acidification, hydrologic flow paths must be considered. Vol. 267; pp. 1639-1647

Schakenbach, J.T.
   Use of calibration gases in the US acid rain program. Accreditation and Quality

Assurance; Jul 01 2001. Description: The United States Acid Rain Program continuous emission monitors (CEMs) have been successful in producing quality-assured data 95% of the time, and in meeting a relative accuracy standard of less than or equal to 10.0% at over 99% of the CEMs in the program. One key reason for this high accuracy is the required use of high quality calibration gases in certification and quality assurance/quality control (QA/QC) tests. An annual QA audit helps ensure high quality calibration gases. Vol. 6; no. 7; pp. 297-301

Scherzer, A.J.; Boerner, R.E.J.
Ambient ozone effects on the ecophysiology of sugar maple (Acer saccharum). American Journal of Botany; Jan 31 1990; pp. 62-63. Description: Sugar maple is among the most widespread and abundant canopy tree species in eastern North America, and is increasing in abundance in the American midwest; yet recent surveys indicate it is declining throughout much of eastern Canada. A number of factors have been cited as causing or contributing to this decline, including both gaseous air pollutants and acidic deposition. The authors hypothesized that ozone has the potential to act as a predisposing factor for sugar maple decline by affecting net carbon gain, carbon allocation, and carbohydrate reserves, resulting in reduced growth and vigor of sugar maple trees.

Schindler, D.W.
The cumulative effects of climate warming and other human stresses on Canadian freshwaters in the new millennium. Canadian Journal of Fisheries and Aquatic Sciences; Jan 01 2001. Description: Climate warming will adversely affect Canadian water quality and water quantity. The magnitude and timing of river flows and lake levels and water renewal times will change. In many regions, wetlands will disappear and water tables will decline. Habitats for cold stenothermic organisms will be reduced in small lakes. Vol. 58; no. 1; pp. 18-29

Schlegel, D.M.; Dippold, D.G.
Planning acid rain compliance: A blend of analytic techniques and engineering judgment. Proceedings of the American Power Conference; Jan 31 1992. Description: This paper reports on the 1990 amendments to the Clean Air act which requires that by the year 2000 electric utilities reduce their emissions of SO_2 to about one half of their 1980 levels. The Clean Air Act (CAA), passed in 1963, allowed the federal government, under certain conditions, to impose air quality rules on the states. Subsequent CAA amendments extended the original CAA legislation to include federal control of certain emission sources, mandated air quality standards, and the requirement that specific compliance measures be used. The 1990 amendments require that coal burning power plants reduce their SO_2 emissions over two Phases. Phase 1 begins in 1995 and continues through 1999; Phase 2 begins in 2000 and continues indefinitely. Vol. 541; pp. 355-359

Schoenberg, S.A.; Benner, R.; Armstrong, A.; Sobecky, P.; Hodson, R.E.
Effects of acid stress on aerobic decomposition of algal and aquatic macrophyte detritus: Direct comparison in a radiocarbon assay. Applied and Environmental Microbiology; Jan 31 1990. Description: Radiolabeled phytoplankton and macrophyte lignocelluloses were incubated at pHs 4 and 7 in water from a naturally acidic freshwater wetland (Okefenokee Swamp; ambient pH, 3.8 to 4.2), a freshwater reservoir (L-Lake; pH 6.7 to 7.2), and a marine marsh (Sapelo Island; pH ~7.8). The data suggest that acidity is an important factor in explaining the lower decomposition rates of algae in Okefenokee Swamp water relative to L-Lake or Sapelo Island water. The decomposition of algal substrate was less sensitive to low pH (~5 to 35% inhibition) than was the decomposition of lignocellulose (~30 to 70% inhibition). These substrate-dependent differences were greater and more consistent in salt marsh than in L-lake incubations. In both freshwater sites, the extent to which

decomposition was suppressed by acidity was greater for green algal substrate than for mixed diatom or blue-green algal (cyanobacteria) substrates. The use of different bases to adjust pH or incubation in a defined saltwater medium had no significant effect on substrate-dependent differences. Although pH differences with lignocellulose were larger in marine incubations, amendment of lake water with marine bacteria or with calcium, known to stabilize exoenzymes in soils, did not magnify the sensitivity of decomposition to acid stress. Vol. 561; pp. 237-244

Schrader, E.L.
Environmental crises of the 21st century: Response and responsibility. Environmental Geology and Water Sciences; Aug 31 1994. Description: This editorial examines the environmental awareness of society today of problems such as ozone depletion, global climate change, oil spills, acid rain, etc. First Schrader discusses the three major components of environmental problems: physical, biological or chemical manifestation of a crisis; who, what caused it; and what is the response to it. Responsibility and response are discussed in greater detail. Vol. 241; pp. 57-58

Schuh, S.A.
Over-breeding: Ethically the ultimate environmental problem. International Journal of Energy-Environment-Economics; Jan 31 1991. Description: The Greenhouse Effect has fuzzy parameters, as do the consequences of acid rain, accidental nuclear fallout, deforestation, even the depletion of oil and natural gas reserves, and other threatening calamities. But the consequences of human over-breeding do not fall within fuzzy parameters. Reliable demographic studies predict a world population by the year 2020 of twice the present four billion or so living human beings. Some of us will see that year. But the population will again have doubled by the year 2090: sixteen billion people. The author suggests in this paper some morally permissible steps that might be taken to circumvent what otherwise is most assuredly an impending world tragedy. We have an ethical obligation to future generations. They have the moral right to a qualitatively fulfilling life, not just on allotted number of years. Some of my suggestions will not be palatable to some readers. But I urge those readers seriously to consider and if possible, hopefully, to propose alternatives. Vol. 12; pp. 105-110

Schuster, P.F.; Reddy, M.M.; Sherwood, S.I.
Effects of acid rain and sulfur dioxide on marble dissolution. Materials Performance; Jan 31 1994. Description: Acid precipitation and the dry deposition of sulfur dioxide (SO_2) accelerate damage to carbonate-stone monuments and building materials. This study identified and quantified environmental damage to a sample of Vermont marble during storms and their preceding dry periods. Results from field experiments indicated the deposition of SO_2 gas to the stone surface during dry periods and a twofold increase in marble dissolution during coincident episodes of low rain rate and decreased rainfall pH. The study is widely applicable to the analysis of carbonate-stone damage at locations affected by acid rain and air pollution. Vol. 331; pp. 76-80

Scrubbers are good
PETC Review (Pittsburgh Energy Technology Center); Jan 31 1992. Description: A flue gas desulfurization (FGD) system, commonly known as a scrubber, is a separate gas cleaning facility installed at the back end of a power plant to remove sulfur dioxide (SO_2). Recent public debate over acid rain control legislation revolved largely around the negative aspects of first-generation scrubbers: their high cost, large size, poor reliability, high energy consumption, and the production of large amounts of scrubber sludge. Vol. 5; pp. 14-19

Segal, M.; Steyn, D.G.
On a strategy for reducing short-range daytime ground level concentrations due to emissions from very tall stacks. Journal of the Air and Waste Management

Association; Feb 28 1990. Description: A procedure with a potential for increasing the duration of tall stack plumes residency within elevated stable layers, outlined schematically. The suggested procedure considers the temporal and spatial behavior of the atmospheric boundary layers (ABL). It takes advantage of the fact that under adverse dispersion conditions, due to relatively shallow ABL, injection of the plume into the capping stable layer above the ABL, may be achieved by a modification of the emission parameters. When successfully applied it will undoubtedly lead to an increase in the time a plume travels within an elevated stable layer, thus ensuring reduced ground-level concentrations as compared to plumes trapped within shallow ABLs. The reduction may be from extremely high concentrations to extremely low concentrations. Application of the strategy is most attractive when the near-stack area is a residential area, while at greater distances from the stack the population is sparse. The technology and engineering involved with increasing emission parameters requires special attention, while the current state-of-the-art tools for forecasting the morning depth of the ABL are quite credible. The proposed strategy has the sole purpose of improving near-field ground-level air quality, in a particular set of meteorological conditions, while retaining the emission rate of gases or even reducing them when compared to normal emission levels.
Vol. 402; pp. 220-223

Selin, Henrik.; Hjelm, Olof.
The role of environmental science and politics in identifying persistent organic pollutants for international regulatory actions. Environmental Reviews; Oct 01 1999. Description: The aim of the present study is to describe and analyze the character of the interplay between environmental science and policy-making in the process of identifying persistent organic pollutants (POPs) for initial inclusion in the POPs Protocol under the Convention on Long-Range Transboundary Air Pollution (CLRTAP). The objective of the CLRTAP POPs Protocol is to control, reduce, or eliminate discharges, emissions, and losses of organic compounds that are toxic, persistent, bioaccumulative, and prone to long-range atmospheric transport and deposition within the CLRTAP region, which covers North America and Europe, including the European region of the former Soviet Union. The empirical materials used were documents underlying decisions and personal observations at seven CLRTAP POPs meetings.
Vol. 7; no. 2; pp. 61-68

Seltzer, R.
U.S. unit to cut sulfur pollution at Polish plant. Chemical and Engineering News; Dec 20 1993. Description: US and Polish officials gathered last month at the Skawina Power Station near Krakow, Poland, to dedicate a showcase pollution control unit that uses advanced US technology and equipment to reduce acid-rain-producing sulfur emissions. The advanced flue gas desulfurization (FGD) unit is viewed as a prototype of efforts by the US Department of Energy to help upgrade pollution control and other operations at power plants in Poland and elsewhere in Eastern Europe. Such projects also offer a large potential market for US companies. Vol. 7151; pp. 17-18

Sersale, R.; Frigione, G; Bonavita, L.
Acid depositions and concrete attack: Main influences. Cement and Concrete Research; Jan 31 1998. Description: The results of an experimental research on the factors responsible to a greater extent for the action of simulated acid precipitations on cement concrete works, both in static and in dynamic conditions, are discussed. The influence of the cement type, the role of calcium hydroxide, the influence of water-cement ratio, and the retard effect on assault, owing to a surface treatment with a water repellent agent, are emphasized.
Vol. 28; no. 1; pp. 19-24

Shah, A.Y; Canter, L.W.
Allowance trading: Correcting the past and looking to the future. Environmental

Professional; Sep 30 1995. Description: Allowance trading is basic to the Title IV acid rain provisions of the 1990 Clean Air Act Amendments (CAAA) in the United States; the provisions seek to achieve a 10-million-ton reduction in annual sulfur dioxide emissions from the electric power utility industry. Allowance trading, a market-based approach, is conceptually similar to the emissions trading policy of the US Environmental Protection Agency (EPA). Vol. 17; no. 3; pp. 193-199

Sharp, R.G.
The Clean Air Act impacts on rail coal. Public Utilities Fortnightly; Mar 01 1991. Description: These factors are examined in this article. In November 1990, President Bush signed the Clean Air Act amendments of 1990 into law. Title IV, concerning acid rain control, calls for a two-phase reduction in power plant sulfur-dioxide emissions, culminating in a nationwide cap after the year 2000. A large part of this reduction will be obtained through substituting low-sulfur coals for the higher-sulfur fuels now used. Most commentators have characterized this legislation as a boon for low-sulfur coal producers and the railroads serving them. Vol. 1275; pp. 26-30

Sharpe, W.E.; DeWalle, D.R.
The effects of acid precipitation runoff episodes on reservoir and tapwater quality in an Appalachian Mountain water supply. Environmental Health Perspectives; Nov 30 1990. Description: The aluminum concentration and Ryznar Index increased and the pH decreased in a small Appalachian water supply reservoir following acid precipitation runoff episodes. Concomitant increases in tapwater aluminum and decreases in tapwater pH were also observed at two homes in the water distribution system. Lead concentrations in the tapwater of one home frequently exceeded recommended levels, although spatial and temporal variation in tapwater copper and lead concentrations was considerable.
Vol. 89; pp. 153-158

Shaw, P.J.; Haan, H. De; Jones, R.I.
The effect of acidification on abiotic interactions of dissolved humic substances, iron and phosphate in epilimnetic water from the HUMEX Lake Skjervatjern Environment International; Jan 31 1992 Description: The responses to pH of abiotic interactions between dissolved humic substances, iron and phosphate were investigated by examining redistributions of $^{55}FeCl_3$ and $^{32}PO_4^{3-}$ added to epilimnetic lakewater from Lake Skjervatjern. The simultaneous movement of ^{55}Fe and ^{32}P to fractions of 10,000-20,000 and > 100,000 Daltons nominal molecular weight, as indicated by Sephadex gel filtration, diminished in response to decreasing pH. Variations in transformations to larger molecular size fractions with incubation time revealed by gel filtration were erratic, but indicated that transformations of added ^{55}Fe and ^{32}P are complete after circa 24 h. Vol. 186; pp. 577-588

Shen, S.; Pepper, G.E; Hassett, J.J; Stucki, J.W.
Acidity and aluminum toxicity caused by iron oxidation around anode bars
Soil Science; Aug 31 1998
Description: Soil acidity and aluminum toxicity are serious environmental problems often found in humid temperate and tropical regions or in areas with acid rain. Iron oxidation in soils can also cause high concentrations of H^+, which, in turn, causes an increase of Al^{3+} in the soil solution. To examine this problem, a study was undertaken to discover the cause of crop damage in crops planted over buried anode bars. Vol. 163; no. 8; pp. 657-664

Shortle, W.C; Smith, K.T; Minocha, R.
Acidic deposition, cation mobilization, and biochemical indicators of stress in healthy red spruce. Journal of Environmental Quality; May 31 1997. Description: Dendrochemical and biochemical markers link stress in apparently healthy red spruce trees (Picea rubens) to acidic deposition. Previous reports related visible damage of trees at high elevations to root and soil processes. In this report, dendrochemical and foliar biochemical markers indicate perturbations in biological processes in

healthy red spruce trees across the northeastern USA. Previous research on the dendrochemistry of red spruce stemwood indicated that under uniform environmental conditions, stemwood concentrations of Ca and Mg decreased with increasing radial distance from the pith. For 9 forest locations, frequency analysis shows 28 and 52% of samples of red spruce stemwood formed in the 1960s are enriched in Ca and Mg, respectively, relative to wood formed prior to and after the 1960's.Vol. 26; no. 3; pp. 871-876

Showman, R.E.; Long, R.P.
Lichen studies along a wet sulfate deposition gradient in Pennsylvania. Bryologist; Jan 31 1992. Description: Lichens were surveyed at four study areas along a wet sulfate deposition gradient in north-central Pennsylvania. This study, performed in September 1988, utilized ten lichen study sites in each of the four study sites was conducted using 7.5' topographic maps in May 1988. Species richness was significantly less in high sulfate deposition areas than in low sulfate deposition areas. Although the causal agents for these differences are not known, it is hypothesized that SO_2 is responsible at least in part. Vol. 952; pp. 166-170

Sigurd Rognerud; Skotvold, Trond.; Fjeld, Eirik.; Norton, Stephen.A.; Anders Hobæk
Concentrations of trace elements in recent and preindustrial sediments from Norwegian and Russian Arctic lakes
Canadian Journal of Fisheries and Aquatic Sciences; Jun 01 1998. Description: Cu, Hg, Ni, Pb, Se, and Zn concentrations in surface and preindustrial freshwater sediments from 66 lakes in the Norwegian and Russian Arctic were used for studying modern atmospheric deposition of these elements. Statistical analysis showed that, after adjusting for the effects of scavenging factors in sediments there were, in general, significantly higher concentrations of Hg and Pb in surface sediments than preindustrial sediments.
Vol. 55; no. 6; pp. 1512-1523

Singh, N.; Yunus, M.; Singh, S.N.; Ahmad, K.J.
Performance of Vicia faba plants in relation to simulated acid rain and/or endosulphan treatment
Bulletin of Environmental Contamination and Toxicology; Feb 28 1992
Description: The increasing human population in India is necessitating the optimum use of cultivable land for increased food production. Eradication and control of pests and pathogens is an essential component of any strategy for increased agricultural production. In this context, the use of pesticides to control the incidence of disease in crops becomes inevitable. Most of the pesticides used for foliar spraying invariably contain surfactants in their formulation, which not only increase the surface wettability but also enhance permeability of the cuticle for more cation infusion/effusion and hence they may make the leaf more susceptible to direct effects of acid rain. To evaluate the validity of this assumption, an experiment using endosulphan, the most commonly used insecticide in India, and acid rain of different pH was conducted on Vicia faba. Vol. 482; pp. 243-248

Sisterson, D.L.; Shannon, J.D.
A comparison of urban and suburban precipitation chemistry
Atmospheric Environment; Jan 31 1990
Description: Precipitation samples at an urban Chicago site and a nearby suburban site were compared in order to examine the influence of emissions within a large urban area on local precipitation chemistry. Precipitation samples were collected from June 1981 to May 1982, initially for events and subsequently weekly, and precipitation- weighted concentrations (PWCs) of the major chemical constituents were calculated from concurrent urban-suburban pairs of samples, stratified according to the estimated mixed-layer wind quadrant. Overall, PWCs at the urban site were higher than those at the suburban site for Ca^{2+}, Mg^{2+}, NH_4^+, NO_3^- and Cl^-; approximately equal for Na^+ and SO_4^{2-}; and lower for H^+. Vol. 243; pp. 389-394

Sjoestroem, J.
Ionic composition and mineral equilibria of acidic groundwater on the west coast of Sweden. Environmental Geology and Water Sciences; Aug 31 1993. Description: The groundwater chemistry of 14 shallow wells and 10 springs in Halland, southwest Sweden, and precipitation have been studied in trilinear diagrams. Ionic strength and saturation index (SI) for selected minerals have been calculated. Five springwaters have similar chemical composition to that of the precipitation, which indicates surficial and rapidly recharged water. The SI of the groundwaters is out of equilibrium (undersaturated) with respect to primary silicates such as mafic minerals, feldspar, K-mica & chlorite, but in equilibrium with solid SiO_2. Vol. 21; no. 4; pp. 219-226

Sklar, S.
A renewable energy strategy Independent Energy; Feb 28 1990 Description: The 1990s are a time of real opportunity for the renewable energy industry. Concerns about world oil prices, global climate change, and the competitiveness of industry in the United States are causing a reevaluation of solar and renewable energy. At the same time the market is expanding on all fronts. In August 1989, SEIA (The Solar Energy Industries Association) presented ten recommendations for policies which might be included as part of President Bush's National Energy Strategy now being prepared by the U.S. Department of Energy. Half of these options would not increase the federal deficit in any way, and the other five options, in aggregate, do not surpass federal support for conventional energy industries. Vol. 202; pp. 49-51

Smith, L.
Sulfate deposition over the Arctic Ocean Ecological Applications; Feb 28 1995 Description: This letter disputes the total sulfate deposition and rainwater pH, as reported in 'Extreme Anthropologenic Loads and the Northern Ecosystem Condition' from November 1993 article. Vol. 6; no. 1; pp. 1

Smith, R.B.
Canada, U.S. fight acid rain Occupational Health and Safety; Nov 30 1991. Description: Short communication. Vol. 6011; pp. 45-46

Sokolov, V.
Managing the global environmental risks in Russia: Missing links and external influences. World Resource Review; Jun 30 1996. Description: Based on analysis of management history of three global environmental issues in Russia--climate change, ozone depletion and acid rains--the author suggests few explanations of failure to build-up the nationwide strategy to manage global risks. Among them are specific factors related to the science-policy relationship in Russia, scientific uncertainties on global changes processes and impacts, etc. Particular attention is given to such internal factors as the monopolization of these issues by the single state agency Hydromet until the late 1980's; the interest of the Soviet military in global atmospheric issues; the absence of any major input from the public or the media; and the manner in which the discussion of these issues was nested within the Soviet government's broader foreign policy agenda.
Vol. 8; no. 2; pp. 198-214

Solomon, B.D
SO_2 allowance trading: What rules apply? Fortnightly; Sep 15 1994
Description: The Acid Rain Program of the Clean Air Act Amendments of 1990 (CAAA) authorized the US Environmental Protection Agency (EPA) to create an innovative system of tradeable SO_2 emission allowances for affected electric utilities. The resulting system is unique in that it requires a partnership between EPA, electric utilities, the Federal Energy Regulatory Commission (FERC), state public utility commissions (PUCs), and state air-pollution control agencies to successfully meet SO_2 reduction requirements. Vol. 132; no. 17; pp. 22-25

Solomon, B.D.; Brick, S.
State regulatory issues in acid rain compliance. Electricity Journal; Mar 31 1992. Description: This article discusses the results of a US EPA workshop for state regulators and commission staff on acid rain compliance concerns. The topics of the article include the results of market-based emissions control, how emissions trading is expected to reduce emissions, public utility commissions approval of compliance plans, the purposes of the workshop, market information, accounting issues, regulatory process and utility planning, multi-state compliance planning, and relationship to other compliance issues. Vol. 52; pp. 20-27

Solomon, B.D.; Rose, K.
Making a market for SO_2 emissions trading Electricity Journal; Jul 31 1992.
Description: Under the innovative, market-based approach to acid rain control included in the Clean Air Act amendments of 1990 (CAAA), sulfur dioxide emission allowances allocated to existing electric utility sources of these emissions can be used by utilities, banked for future use, or sold or traded to other users. Most power plants that burn fossil fuels will need to obtain an adequate supply of allowances from the market of EPA-sponsored auctions to cover their future emissions. This article addresses the respective roles of regulators and the private sector in facilitating a market for SO_2 emission allowances. In previous work, the authors have argued that state public utility commissions should seize the opportunity to encourage utilities to facilitate the allowance market. Yet it is the nature of new markets that many potential participants (including regulators) are risk-averse and wait for others to make the first move. Taken to the extreme, such behavior is a prescription for failure.
Vol. 56; pp. 58-66

Sorini, S.S.
Development and validation of a standard test method for sequential batch extraction of waste with acidic extraction fluid Journal of Testing and Evaluation; Mar 31 1994. Description: A project was conducted to develop and validate a sequential batch extraction method using a dilute acid solution as the extraction fluid. The method was developed by ASTM Task Group D-34.02.01 on Waste Leaching Techniques, and calls for the pH of the extraction fluid to reflect the pH of acidic precipitation in the geographic region where the material being tested is to be disposed. A collaborative study of the method was conducted to determine the multiple- laboratory and single-operator precision of the extraction procedure when applied to two different waste material using two different pH values
Vol. 222; pp. 168-174

Spainhoward, D.A; Schultz, D.E.
Mix coal switching with emission trading Fortnightly; May 01 1995
Description: With the Clean Air Act Amendments of 1990 (CAAA) come many complex decisions for electric utilities. By now the majority of utilities have decided how they will comply with the clean air guidelines and acid rain program limits, at least for Phase I. But for those utilities that have selected coal switching as the preferred method of complying with the law, the task gets more complicated. This article describes a strategy for developing least-cost compliance methods by integrating the cost of emission allowances in the coal bid evaluation process.
Vol. 133; no. 9; pp. 18-19

Spengler, J.D.; Brauer, M.; Koutrakis, P.
Acid air and health. Environmental Science and Technology; Jul 31 1990. Description: The health effects of acid air pollution are a national and an international concern. Research programs often are divided between exposure and toxicological assessments. Ideally, the exposure and toxicity components of acidic air pollution risk assessment should be linked at key junctures. The understanding of particle size, delivery system, inter-subject variability in doses and susceptibility, exposure chamber reaction, and inhalation chemistry may help to reconcile the results of controlled human exposure. These

studies currently contribute more uncertainty to the overall determination of a relationship between atmospheric acidity and health effects than do current techniques to characterize exposure to atmospheric acidity. Data are presented on the daily hydrogen ion concentrations in six US cities as well as concentration ranges of SO_4^{2-} and H^+ (as HSO_4) in selected North American cities. A summary of epidemiologic studies of acid aerosol exposures is presented.
Vol. 247; pp. 946-956

Spiegel, R.J.; Thorneloe, S.A.; Trocciola, J.C.; Preston, J.L.
Fuel cell operation on anaerobic digester gas: Conceptual design and assessment
Waste Management; Nov 01 1999
Description: The conceptual design of a fuel cell (FC) system for operation on anaerobic digester gas (ADG) is described and its economic and environmental feasibility is projected. ADG is produced at wastewater treatment plants during the process of treating sewage anaerobically to reduce solids. The economic feasibility study shows the fuel cell is economical where plant electricity costs are 5 [cents]/kW h or higher, based on entry level fuel cell costs of $3,000/kW. FCs are one of the cleanest energy technologies available, and the widespread use of this concept should result in a significant reduction in global warming gas and acid rain air emissions. Additionally, technology evaluation focused on improving a commercial phosphoric acid FC power plant operation on ADG is described. Vol. 19; no. 6; pp. pp.389-399

Spills, drills, and accountability
Amicus Journal; Jan 31 1993
Description: NRDC seeks preventive approaches to oil pollution on U.S. coasts. The recent oil spills in Spain and Scotland have highlighted a fact too easy to forget in a society that uses petroleum every minute of every day: oil is profoundly toxic. One tiny drop on a bald eagle's egg has been known to kill the embryo inside. Every activity involving oil-drilling for it, piping it, shipping it-poses risks that must be taken with utmost caution. Moreover, oil production is highly polluting. It emits substantial air pollution, such as nitrogen oxides that can form smog and acid rain. The wells bring up great quantities of toxic waste: solids, liquids and sludges often contaminated by oil, toxic metals, or even radioactivity. Vol. 15; no. 1; pp. 3-4

Srivastava, R.K.; Jozewicz, W.
Flue gas desulfurization: the state of the art
Journal of the Air and Waste Management Association; Dec 01 2001. Description: Coal-fired electricity-generating plants may use SO_2 scrubbers to meet the requirements of phase II of the Acid Rain SO_2 Reduction Program. Additionally, the use of scrubbers can result in reduction of Hg and other emissions from combustion sources. It is timely, therefore, to examine the current status of SO_2 scrubbing technologies. This paper present a comprehensive review of the state of the art in flue gas desulfurization (FGD) technologies for coal fired boilers. 30 refs., Vol. 51; no. 12; pp. 1676-1688

Stam, A.C.; Mitchell, M.J.; Krouse, H.R.; Kahl, J.S.
Stable sulfur isotopes of sulfate in precipitation and stream solutions in a northern hardwood watershed
Water Resources Research; Jan 31 1992
Description: Stable S isotopes of SO_4^{2-} in precipitation, throughfall, and stream water solutions in a northern hardwood watershed (Bear Brook Watershed, Maine) were examined to determine sources of stream SO_4^{2-} and to identify watershed processes that may affect atmospherically deposited SO_4^{2-} prior to reaching the streams. Vol. 281; pp. 231-236

Stauffer, R.E.; Wittchen, B.D.
Effects of silicate weathering on water chemistry in forested, upland, felsic terrane of the USA. Geochimica et Cosmochimica Acta; Nov 30 1991. Description: The authors use data from the US EPA National Surface Water Survey (NSWS), the USGS Bench-Mark Station monitoring program, and the National Acid Deposition Program (NADP) to evaluate the role of

weathering in supplying base cations to surface waters in forested, upland, felsic terrane of the northeastern, northcentral, and northwestern (Idaho batholith) US. Multivariate regression reveals differential effects of discharge on individual base cations and silica, but no secular trend in the Ca/Na denudation rate over 24 yr (1965-1988) for the Wild River catchment in the White Mountains. Because the turnover time for Na in the soil-exchange complex is only ca. 1.5 yr, the long-term behavior of the ratios Ca/Na and Si/Na in waters leaving this catchment indicates that weathering is compensating for base cation export. Vol. 5511; pp. 3253-3271

Stauffer, R.E.; Wittchen, B.D.
Thermal and trophic stability of deeper Maine lakes in granite waterhsheds implacted by acid deposition
Water Resources Research; Sep 30 1990
Description: Acid deposition can lead to lake and watershed acidification, increases in lake transparency, and reduction in thermal stability and hypolimnetic oxygen deficits. On the basis of lake surveys during August-September 1985, we determined to what extent the deeper (maximum depth $z_m\{gt\}17$ m) Maine lakes in acid-sensitive granitic watersheds have registered changes in temperature and oxygen stratification, as compared to 1938-1942, when G.P. Cooper performed the earliest scientific surveys of the state's lakes. After correcting for small but geographically consistent interannual differences in summer hypolimnetic temperatures related to spring turnover, and weather-dependent differences in mixed layer depth, there has been no significant change in thermal stratification in these Maine lakes over approximately 43 years. Vol. 269; pp. 2143-2151

Stein, H.
Utilities find it difficult to meet uncertain NO_x control requirements. Electric Light and Power (Boston); Mar 31 1994.
Description: US electric utilities, and eventually all utilities, will have to reduce their emissions of what are popularly and politically classified as "harmful" emissions. But it is not proving easy to hit the moving targets being produced by governmental bodies. One of the high-profile targets is the emission of oxides of nitrogen (NO_x), which are considered among the most prevalent and also are classed as harmful precursors of both ozone and acid rain formation. Different technologies for reducing NO_x emissions are described. Vol. 723; pp. 37-39

Stigliani, W.M.; Shaw, R.W.
Energy use and acid deposition: The view from Europe. Annual Review of Energy; Jan 31 1990. Description: This paper reviews the chemistry behind acid deposition patterns and environmental effects. Energy use is correlated to acid deposition. Much of this paper is focused on Europe, and the Netherlands and Poland. Vol. 15; pp. 201-216

Stockwell, W.R.; Middleton, P.; Chang, J.S.; Tang, Xiaoyan
The second generation regional acid deposition model chemical mechanism for regional air quality modeling. Journal of Geophysical Research; Sep 20 1990. Description: A state-of-the-art gas phase chemical mechanism for modeling atmospheric chemistry on a regional scale is presented. The second generation Regional Acid Deposition Model (RADM2) gas phase chemical mechanism, like its predecessor RADM1, is highly nonlinear, since predicted ozone, sulfate, nitric acid and hydrogen peroxide concentrations are complicated functions of NO_x and nonmethane hydrocarbon concentrations. The RADM2 chemical mechanism is an upgrade of RADM1 in that (1) three classes of higher alkanes are used instead of one, (2) a more detailed treatment of aromatic chemistry is used, (3) the two higher alkene classes now represent internal and terminal alkenes, (4) ketones and dicarbonyl species are treated as classes distinct from aldehydes, (5) isoprene is now included as an explicit species, and (6) there is a more detailed treatment of peroxy radical-peroxy radical reactions. As a result of these improvements the RADM2 mechanism

simulates the concentrations of peroxyacetyl nitrate, HNO3, and H_2O_2 under a wide variety of environmental conditions. Comparisons of RADM2 mechanism with the RADM1 mechanism predictions and selected environmental chamber experimental results indicate that for typical atmospheric conditions, both mechanisms reliably predict O_3, sulfate and nitric acid concentrations. The RADM2 mechanism gives lower and presumably more realistic predictions of H_2O_2 because of its more detailed treatment of peroxy radical-peroxy radical reactions.
Vol. 9510; pp. 16,343-16,367

Strandberg, H; Johansson, L.G.
Some aspects of the atmospheric corrosion of copper in the presence of sodium chloride. Journal of the Electrochemical Society; Apr 30 1998. Description: The effect of NaCl in combination with O_3 and SO_2 on the atmospheric corrosion of copper was investigated. Corrosion products formed after 4 weeks exposure were characterized qualitatively by X-ray diffraction and quantitatively by gravimetry and ion chromatography of leaching solutions. Studies of SO_2 deposition and O_3 consumption were performed using on-line gas analysis.
Vol. 145; no. 4; pp. 1093-1100

Stricker, G.D.
Alaska has 4.0 trillion tons of low-sulfur coal: Is there a future for this resource AAPG Bulletin (American Association of Petroleum Geologists); May 31 1990 Description: The demand for and use of low-sulfur coal may increase because of concern with acid rain. Alaska's low-sulfur coal resources can only be described as enormous: 4.0 trillion tons of hypothetical onshore coal. Mean total sulfur content is 0.34% (range 0.06-6.6%, n = 262) with a mean apparent rank of subbituminous B. There are 50 coal fields in Alaska; the bulk of the resources are in six major fields or regions: Nenana, Cook Inlet, Matanuska, Chignik-Herendeen Bay, North Slope, and Bering River. For comparison, Carboniferous coals in the Appalachian region and Interior Province have a mean total sulfur content of 2.3% (range 0.1-19.0%, n = 5,497) with a mean apparent rank of high-volatile A bituminous coal, and Rocky Mountain and northern Great Plains Cretaceous and Tertiary coals have a mean total sulfur content of 0.86% (range 0.02-19.0%, n = 2,754) with a mean apparent rank of subbituminous B. Alaskan coal has two-fifths the total sulfur of western US coals and one-sixth that of Carboniferous US coals. Even though Alaska has large resources of low-sulfur coal, these resources have not been developed because of (1) remote locations and little infrastructure, (2) inhospitable climate, and (3) long distances to potential markets. These resources will not be used in the near future unless there are some major, and probably violent, changes in the world energy picture. Vol. 745; pp. 772

Stubbs, C.S.; Homola, R.H.
Not rare. But, endangered Elemental profiles of three corticolous lichen species on red spruce in Maine. American Journal of Botany; Jan 31 1990; pp. 4. Description: Usnea subfloridana Stirton, Platismatia glauca (L.) Club. and Club., and Hypogymnia physodes (L.) Nyl. are lichen species moderately to highly sensitive to air pollutants, including acid deposition and ozone. Some researchers have attributed depauperate populations and local extinctions of these species to poor air quality. Since 1985, areas of Maine annually experienced mean summer rain and fog events of pH 4.5 or lower and ozone levels above national standards. Given this possible threat to these and other pollution sensitive species, baseline elemental analyses for Ca, K, P, Mg, Al, B, Fe, Cu, Mn, Zn, N, S, Na, and Pb were performed in 1986 on coastal and inland populations on Picea rubens L. Elemental analyses were again performed on nontransplanted and transplanted lichens from the same populations in 1988.

Sugg, P.M; Kuperman, R.G; Loucks, O.L.
Assessing biogeographic patterns in the changes in soil invertebrate biodiversity due to acidic deposition. Bulletin of the Ecological Society of America; Sep 30

1995. Description: We are studying the response of soil faunal communities to a gradient in acidic deposition across midwestern hardwood forests. We have documented a pattern of population decrease and species loss for soil invertebrates along the acidification gradient. We now ask the following question: When confronted with apparent diversity changes along a region-wide pollution gradient, how can one assess the possibility of natural biogeographic gradients accounting for the pattern? As a first approximation, we use published range maps from taxonomic monographs to determine the percent of the regional fauna with ranges encompassing each site. Vol. 76; no. 3; pp. 391

Sullivan, T.J.; Kugler, D.L.; Johnson, C.B.; Rosenbaum, B.J.; Small, M.J.; Landers, D.H.; Overton, W.S.; Kretser, W.A.; Gallagher, J.
Variation in Adirondack, New York, lakewater chemistry as function of surface area. Water Resources Bulletin; Feb 28 1990. Description: Data from a recent survey conducted by the Adirondack Lake Survey Corporation were used to evaluate the influence of lake surface area on the acid-base status of lakes in Adirondack State Park, New York. Acid neutralizing capacity (ANC) in the small lakes (<4 ha) occurred more frequently at extreme values (>200, <0 {mu}eq L^{-1}), whereas larger lakes tended to be intermediate in ANC. Consequently, acidic (ANC {le} 0) and low-pH lakes were typically small. The small lakes also exhibited lower Ca^{2+} concentration and higher dissolved organic carbon than did larger lakes. Lakes {ge} 4 ha were only half as likely to be acidic as were lakes {ge} 1 ha in area. These data illustrate the dependence of lake chemistry on lake surface area and the importance of the lower lake area limit for a statistical survey of lakewater chemistry. Vol. 261; pp. 167-176

Suomela, Janne.; Neuvonen, Seppo.
Effects of long-term simulated acid rain on suitability of mountain birch for Epirrita autumnata (Geometridae). Canadian Journal of Forest Research; Feb 01 1997. Description: We studied the effects of simulated acid rain (sulphuric and nitric acids) on foliage quality of mountain birch (Betula pubescens ssp. czerepanovii (Orlova) Hämet-Ahti) for the geometrid Epirrita autumnata Bkh., which is the most destructive defoliator of mountain birch forests. The study was conducted in northern Finland, where ambient pollution is low but environmental conditions are otherwise harsh. Sulphuric acid, nitric acid, or their combination had no effect on relative growth rate of E. autumnata larvae. Vol. 27; no. 2; pp. 248-256

Svensson, J.E; Johansson, L.G.
The synergistic effect of hydrogen sulfide and nitrogen dioxide on the atmospheric corrosion of zinc. Journal of the Electrochemical Society; Jan 31 1996. Description: A laboratory study of the effect of sub-ppm levels of H_2S and NO_2 on the atmospheric corrosion of zinc in humid air is reported. Each sample was exposed individually to a synthetic atmosphere with careful control of pollutant concentrations, relative humidity, and flow conditions. Corrosion products were analyzed by grazing angle X-ray diffraction and X-ray photoelectron spectroscopy. Ion chromatography was employed to identify water soluble anions. The interaction of the pollutants with zinc metal was studied using trace gas analysis in real time. Vol. 143; no. 1; pp. 51-58

Tabors, R.D.; Monroe, B.L. III
Planning for future uncertainties in electric power generation: An analysis of transitional strategies for reduction of carbon and sulfur emissions. IEEE Transactions on Power Systems (Institute of Electrical and Electronics Engineers); Nov 30 1991. Description: The objective of this paper is to identify strategies for the U.S. electric utility industry for reduction of both acid rain producing and global warming gasses. The research used the EPRI Electric Generation Expansion Analysis System (EGEAS) utility optimization/simulation modeling structure and the EPRI developed regional utilities.

It focuses on the North East and East Central region of the U.S. Strategies identified were fuel switching -- predominantly between coal and natural gas, mandated emission limits, and a carbon tax. Vol. 64; pp. 1500-1507

Tahvonen, O.; Kaitala, V.; Pohjola, M.
A Finnish-Soviet acid rain game: Noncooperative equilibria, cost efficiency, and sulfur agreements. Journal of Environmental Economics and Management; Jan 31 1993. Description: This study analyzes cost effectiveness in environmental cooperation between Finland and the Soviet Union. It is assumed that the aim of both countries is to attain a given target deposition level at minimum possible sulfur abatement costs. Cost-effective cooperation is compared to noncooperative equilibrium and to the agreement on sulfur emissions between these two countries. Vol. 241; pp. 87-100

Tajchman, S.J.; Kosuri, S.R.; Zeleznik, J.D.
Spatial variation in acidic deposition in an appalachian forest. Northern Journal of Applied Forestry; Mar 31 1993. Description: Precipitation is the main source of water for the forest. Gross precipitation reaching the forest canopy is partitioned among interception, stem flow, and through fall. While the intercepted fraction of precipitation evaporates into the atmosphere, stem flow and through fall reach the forest floor, but their chemistry is as a rule different from that of gross precipitation. This modification is due to the enrichment of stem flow and through fall with substances of dry deposition washed off from the vegetation surface. Vol. 101; pp. 41-43

Takamatsu, T.; Sase, H.; Takada, J.
Some physiological properties of *Cryptomeria japonica* leaves from Kanto, Japan: potential factors causing tree decline. Canadian Journal of Forest Research; Apr 01 2001. Description: Japanese cedar (*Cryptomeria japonica* D. Don) has been declining in urban areas of Japan. We examined if the decline was associated with physiological deterioration of leaves and resulting water stress. Leaves from three locations (severe decline, slight decline, and healthy) were analyzed for minimum transpiration rates (MT), amounts of epicuticular wax (EW), contact angles (CA), fractions of unhealthy stomata (US), cuticular thickness, and leaching of elements (LE). Anthropogenic elements (e.g., antimony (Sb)) in aerosols on the leaves were also analyzed by neutron activation analysis.
Vol. 31; no. 4; pp. 663-672

Ten utilities receive acid rain bonus allowances from EPA. Energy Efficiency Journal; Jan 31 1995. Description: The United States Environmental Protection Agency (EPA) recently awarded 1,349 acid rain bonus allowances to ten utilities for energy efficiency and renewable energy measures. An allowance licensesthee emission of one ton of sulfur dioxide. A limited number of allowances are allocated to utilities to ensure that emissions will be cut to less than 9 million tons per year.
Vol. 3; no. 2; pp. 4

Tidblad, J.; Leygraf, C.; Kucera, V.
Acid deposition effects on materials: Evaluation of nickel after four years of exposure. Journal of the Electrochemical Society; Jul 31 1993. Description: The atmospheric corrosion of nickel after 4 years of exposure inside a sheltered box has been investigated. Quantitative evaluation of corrosion attack resulted in a linear relation between the weight increase of nickel and the sulfur dioxide concentration. Analysis of corrosion products by X-ray photoelectron spectroscopy combined with diffuse reflectance infrared Fourier transform spectroscopy and X- ray powder diffraction agreed with previous conclusions and suggested the initial formation of an amorphous basic nickel sulfate with less protective ability and subsequent formation of a crystalline basic nickel sulfate with higher protective ability. A carbonate was also detected, more abundant at lower weight increases.
Vol. 1407; pp. 1912-1916

Tidblad, J; Leygraf, C.
Atmospheric corrosion effects of SO_2 and NO_2: A comparison of laboratory and field-exposed copper. Journal of the Electrochemical Society; Mar 31 1995. Description: Laboratory exposures of copper have been performed at exposure conditions comparable to those in the UN ECE exposure program with respect to air flow conditions, relative humidity, and concentration of the gaseous pollutants sulfur dioxide and nitrogen dioxide. Extrapolation of the weight increases in the laboratory experiments match well those obtained at the test sites with high sulfur dioxide and nitrogen dioxide pollution levels. At these sites, sulfate and nitrate were the dominating surface constituents, as in the laboratory exposures.
Vol. 142; no. 3; pp. 749-756

Todd, D.M.
Clean coal technologies for gas turbines Turbomachinery International; Jan 31 1992 Description: Currently in the USA, many states include, or have orders pending to include, environmental externality costs as part of new generation siting evaluation. The externality values for environmental evaluations vary from state to state and are generally developed from marginal cost assessments. A 1989 Pace University study derived representative externality values ($/lb) from direct cost estimation techniques and are shown for both acid rain and greenhouse gas emissions. Using the Pace values, externality assessments were made for each of the technology plants, and illustrate the cumulative magnitude of the externality valuation process. Based on current and future performance projections, Integrated Gasification Combined-Cycle (IGCC) plants compare favorably against direct coal fired alternatives, and will continue to trend towards conventional gas fired combined-cycle plant externality values with further near-term technology advances. Vol. 332; pp. 20-27

Tombach, I.
Emissions trading -- Market-based approaches offer pollution control incentives. Environmental Solutions; Jun 30 1994. Description: In the last several years, market-based" strategies for achieving air quality goals have joined the traditional command-and-control" approach to air pollution management. The premise behind market approaches is that the right" to emit air pollutants provided by a permit has monetary value. A market-based approach provides facility operators with incentives to take advantage of the monetary value associated with reducing emissions below permitted levels.
Vol. 76; pp. 50-55

Tonnessen, K.A.
Emerald Lake Watershed study: Introduction and site description Water Resources Research; Jul 31 1991 Description: The Emerald Lake Watershed study was organized to investigate the effects of acidic deposition on high-elevation watersheds and surface waters of the Sierra Nevada, California. Some of the results of this comprehensive study of aquatic and terrestrial ecosystems at a small, headwater basin are presented in four papers in this series. The watershed study site is in Sequoia National Park, on the western slope of the Sierra Nevada.
Vol. 277; pp. 1537-1539

Traina, J.; Foreman, J.
The long term accuracy and maintainability of an ultrasonic flow monitor on an FGD application. Proceedings of the American Power Conference; Jan 31 1992. Description: EPA's proposed Acid Rain regulations (40CFR75) will, within just a few years, require nearly every utility in the United States to monitor mass emissions (pounds/hour) of SO_2 to standards of accuracy and accountability similar to those now in effect for gas concentration analyzers. This paper describes a monitoring technique - ultrasonic transit time - which has demonstrated the capability of meeting those accuracy and drift standards. Vol. 541; pp. 695-699

Tsujita, Cameron.J.
   The significance of multiple causes and coincidence in the geological record: from clam clusters to Cretaceous catastrophe Canadian Journal of Earth Sciences; Feb 01 2001. Description: Specific causes of unusual events recorded in the geological record are commonly difficult to distinguish and isolate; in some instances, event strata contain features that cannot be explained by a single causal mechanism. Unicausal hypotheses, when applied to complex problems, can lead to the misidentification, misinterpretation, and force-fitting of observations ("great expectations syndrome"). The close timing or temporal overlap of significant events, although statistically improbable on short time scales, becomes possible on long time scales. Vol. 38; no. 2; pp. 271-292

Turner, R.S.; Dale, V.H.; Olson, R.J.
   Ecology and the integrated assessment process. Bulletin of the Ecological Society of America; Jun 30 1994. Description: Integrated assessment is a process in which technical, social, environmental, and economic consequences of alternative decision scenarios are evaluated. It requires evaluation and accomodation of multiple, sometimes opposing or conflicting, points of view or models of the way things work. The goal is not the development or imposition of a single unified or integrated world model, no matter how comprehensive or sophisticated, but rather the building of a consensus among stakeholders about the direction, magnitude, and uncertainties of the societal and ecological effects of alternative decisions. Vol. 752; pp. 233

US industrial SO_2 emissions expected to remain steady. Air Pollution Consultant; Mar 31 1996. Description: Although reductions in U.S. sulfur dioxide (SO_2) emissions are being achieved primarily through emission reductions at utilities, two other significant SO_2 source groups are also being monitored: industrial sources and diesel fuel users. As part of its actions associated with acid rain prevention, congress mandated EPA to (1) inventory national SO_2 emissions from industrial sources, (2) predict the trend in industrial emissions expected over the next 20 years, and (3) estimate emission reductions achieved through diesel fuel desulfurization regulations. In response to the mandate, EPA has released a report entitled National Annual Industrial Sulfur Dioxide Emission Trends 1995-2015: Report to congress. The highlights of the report are presented here. Vol. 6; no. 2; pp. 1.27-1.28

Vaghjiani, G.L.; Ravishankara, A.R.
   Photodissociation of H_2O_2 and CH_3OOH at 248 nm and 298 K: Quantum yields for OH, $O(^3P)$ and $H(^2S)$. Journal of Chemical Physics; Jan 15 1990. Description: The quantum yields of the products, $OH X^2\pi$), $O(^3P)$ (plus $O(^1D)$) and $H(^2S)$, in the photolysis of H_2O_2 and CH_3OOH at 248 nm and 298 K have been measured. OH was directly observed by laser-induced fluorescence while the atomic species were detected by cw-resonance fluorescence. All quantum yield measurements were made using relative methods.
   Vol. 922; pp. 996-1003

Vogt, R.D.; Seip, H.M.; Ranneklev, S.
   Soil and soil water studies at the HUMEX site. Environment International; Jan 31 1992. Description: Changes to natural organic compounds by acid deposition and subsequent effects on Al mobilization are not well understood. The HUMEX catchment-scale acidification experiment in western Norway offers a unique possibility for an integrated assessment of these interactions. In this report, the soil and soil water chemical data from the HUMEX site, from before and after the onset of experimental acidification, are used to characterize the catchment.
   Vol. 186; pp. 555-564

Vong, R.J.; Guttorp, P.
   Co-occurrence of ozone and acidic cloudwater in high-elevation forests Environmental Science and Technology; Jul 31 1991. Description: A chemical climatology for high-elevation forests was estimated from ozone and cloudwater acidity data collected in the eastern US.

Besides frequent ozone-only and pH-only single-pollutant episodes, both simultaneous and sequential co-occurrence of ozone and acidic cloud-water were observed a few times each month above cloud base. Co-occurrence was observed more frequently at two southern sites than at two northern sites. This co-occurrence represents a multiple chemical stress whose biological implications can now for the first time be studied in controlled exposure of trees. Vol. 257; pp. 1325-1329

Wagman, J.D.; Downey, D.
Limestone treatment of acidified streams Virginia Journal of Science; Jan 31 1991 Description: The goal of this study is to evaluate the mitigation of the acidification of streamwater through single-point application of limestone. Acid rain is primarily formed when nitrogen oxides (NO_x) and sulfur oxides (SO_x) emitted from the burning of fossil fuels react in the atmosphere to form nitric and sulfuric acids, respectively. These acids return to earth where they enter streams through rain runoff and underground water displacement. The sulfate ion reduces the stream's buffering capacity by replacing the natural carbonate and bicarbonate. Limestone is calcium carbonate ($CaCO_3$) which dissolves sufficiently in streams to provide supplemental carbonate ions to boost buffering capacity.
Vol. 422; pp. 205

Wagner, G.H.; Steele, K.F.; Peden, M.E.
Dew and frost chemistry at a midcontinent site, United States. Journal of Geophysical Research; Dec 20 1992. Description: Little national effort is being devoted to appraising the importance of dew in the research on acid rain and atmospheric pollutants. Because dew lingers directly on plants and is perhaps more concentrated than rain, especially during its evaporation, it may overshadow certain rain effects which work mainly through the soil. From July 1989 to July 1990 a total of 98 dew and 9 frost samples were collected at the University of Arkansas Agricultural Experiment Station, Fayetteville.
Vol. 9718; pp. 20591-20597

Walker, R.F.; McLaughlin, S.B.
Growth and xylem water potential of white oak and loblolly pine seedlings as affected by simulated acidic rain. American Midland Naturalist; Jan 31 1993. Description: Effects of simulated acidic rainfall on the growth and water relations of white oak (Quercus albya L.) and loblolly pin (Pinus taeda L.) seedlings grown under two fertility regimes were examined. Seedling of each species grown in a loam soil were exposed to two simulated rains per week of pH 4.8, 4.2 or 3.6 for 26 wk. High and low fertility regimes were imposed by monthly application of one-half and one-quarter concentration, respectively, of Hoagland's solution No. 2. Diameter growth of both species was reduced by exposure to rains of the higher acidities regardless of fertility treatment, and seedlings that received pH 4.2 and 3.6 rains also exhibited greater foliar chlorosis and necrosis than those that received rains of pH 4.8.
Vol. 1291; pp. 26-34

Walse, Charlotta.; Berg, Björn.; Sverdrup, Harald.
Review and synthesis of experimental data on organic and acidity. Environmental Reviews; Jan 01 1998. Description: A review and synthesis of experimental decomposition data was performed with the objective of finding parameter values for a decomposition model. Experimental data were retrieved from the literature and included data on mass loss rates, nitrogen mineralization rates, carbon dioxide evolution rates, and growth rates of bacteria and fungi. Environmental variables included in the synthesis were air temperature, soil moisture, and soil acidity (concentration of H^+ and Al^{3+} in soil solution). Vol. 6; no. 1; pp. 25-40

Wang, J.H; Shih, H.C.; Wei, F.I.
Electrochemical studies of the corrosion behavior of carbon and weathering steels in alternating wet/dry environments with sulfur dioxide gas. Corrosion (Houston); Aug 31 1996. Description: Electrochemical impedance techniques were used to investigate the corrosion behavior of

carbon steel (CS) and weathering steel (WS) in sulfur dioxide (SO_2)-containing environments. Impedance measurements were conducted in a modified three-electrode electrochemical cell covered by a thin electrolyte layer during the wet/dry period. Results showed WS was more resistant to SO_2-induced atmospheric corrosion than CS. Three forms of impedance spectra were observed, depending upon exposure period. Accordingly, three impedance models were proposed to explain the characteristic impedance data and corrosion behaviors in different stages of exposure. The proposed models and equivalent circuits produced good agreement with experimental impedance data. Vol. 52; no. 8; p. 600-608

Wang, X.; Chen, L.; Yoshimura, N.
Erosion by acid rain, accelerating the tracking of polystyrene insulating material Journal of Physics D: Applied Physics; May 07 2000. Description: Because outdoor insulating materials in service are subjected to numerous wet and dry cycles, it is necessary to establish their performance in acid rain. The erosion effect of acid rain on atactic polystyrene insulating material is investigated using accelerated ageing by artificial acid rain. The degradation mechanisms of material structure and tracking resistance are discussed. Vol. 33; no. 9; pp. 1117-1127

Warkentin, D
Utilities swap pollution credits. Electric Light and Power (Boston); Mar 31 1995. Description: In an innovative plan to reduce acid rain and greenhouse gas emissions, Niagara Mohawk Power Corp. (NMPC) will transfer 1.75 million tons of CO_2 reductions to Arizona Public Service Co. (APS) in return for 25,000 tons of SO_2 allowances. The two companies agreed to the plan - the first-ever interpollutant credit swap - late in 1994. NMPC will donate the APS-supplied SO_2 allowances to a non-profit environmental organization. The allowances will be permanently retired and not released into the atmosphere. NMPC will use the tax benefit resulting from the donation, which is expected to be about $1 million, to fund energy efficiency programs or energy supply networks. Vol. 73; no. 3; pp. 1, 9

Watmough, Shaun.A.
An evaluation of the use of dendrochemical analyses in environmental monitoring. Environmental Reviews; Mar 01 1997. Description: Dendrochemical techniques have been used to monitor historical changes in soil and atmospheric chemistry since the early 1970s. The development of dendrochemistry in environmental monitoring was prompted by early studies which reported that changes in Pb deposition along roadsides and in industrial areas were reflected by changes in the Pb content of tree rings. Vol. 5; no. 3; pp. 181-201

Weinstein, D.
Complying with the 1990 Amendments: Managing risks and preparing for prudence reviews. Natural Resources and Environment; Jan 31 1992. Description: The 1990 Clean Air Act Amendments will affect the costs, competitive position, and profitability of any company that must comply with Title I (nonattainment), Title II (mobile sources), Title II (air toxics), or Title IV (acid Rain). This article describes ways lawyers can actively help their clients comply with the Amendments by implementing a strategic risk-management planning process. This usually is a multidiciplinary approach and involves legal counsel as well as economic and engineering advice. The result will be a compliance plan that not only minimizes costs but also manages risks. Vol. 7; no. 2; pp. 31-33, 56-57

Weller, R.; Schrems, O.
H_2O_2 in the marine troposphere and seawater of the Atlantic Ocean (48°N-63°S). Geophysical Research Letters (American Geophysical Union); Jan 22 1993. Description: Concentrations of H_2O_2 in gas phase and seawater have been measured in pristine regions of the Atlantic Ocean during the RV Polarstern expedition ANT X-1 from 11/15/91 to 01/02/92. A broad maximum of gaseous H_2O_2 mixing

ratio in the troposphere was found between the tropic of cancer and the tropic of capricornus with peak values around 1.8 ppbv. The observed ratio of organic peroxides/total amount of peroxides was 0.2-0.35, in contrast to the remarkably lower ratio of 0.05-0.35, in contrast to the remarkably lower ratio of 0.05-0.10 measured in the continental troposphere by other groups. Vol. 202; pp. 125-128

Weng, Chengyu.; Jackson, Stephen.T.
Species differentiation of North American spruce (*Picea*) based on morphological and anatomical characteristics of needles Canadian Journal of Botany; Nov 01 2000 Description: Differentiation of most North American spruce (*Picea*) species can be done based on needle morphology and anatomy. *Picea breweriana* S. Watson, *Picea chihuahuana* Martìnez, *Picea mariana* (Mill.) BSP, *Picea martinezii* Patterson, and *Picea rubens* Sarg. needles have two continuous resin ducts extending from near the base to near the tip. *Picea engelmannii* Parry ex Engelm., *Picea glauca* (Moench) Voss, *Picea pungens* Engelm., *Picea mexicana* Martìnez, and *Picea sitchensis* (Bong.) Carr. needles have variable numbers of short, intermittent resin ducts or sacs.
Vol. 78; no. 11; pp. 1367-1383

Westenbarger, D.; Frisvold, F.
Air quality and land productivity in the northeastern United States, 1980-85 American Journal of Agricultural Economics; Dec 31 1992. Description: This study estimates the impact of ozone pollution and acid rain on agricultural land productivity. Sulfate depositions and ozone reduce productivity, while nitrate depositions increase it. The countervailing effects of sulfate and nitrate depositions cancel each other out. The net effect of acid depositions is negligible over the sample region. Vol. 745; pp. 1284

White, D.F.; Fornes, R.E.; Gilbert, R.D.; Speer, J.A.
Investigation of the formation of physical damage on automotive finishes due to acidic reagent exposure. Journal of Applied Polymer Science; Oct 15 1993.
Description: Automotive paints with clear-coat surfaces can be physically damaged by exposure to acidic reagents produced in a smog chamber designed to reproduce real environmental conditions. Visual and reflectance microscopy observations show that deposition of material formed from the reaction of the clear coat and the reagent drop occurs on the paint surface after the drop evaporates to a critical size, with the greatest deposition occurring at the edge of the drop. This type of deposition suggests a free-energy minimization process favoring the formation of stable nuclei at the reagent drop edge. Vol. 503; pp. 541-549

White, J.P.; Mitnick, S.A.
Clean air land mine: Continuous monitoring. Public Utilities Fortnightly; Dec 01 1992. Description: When the Clean Air Act Amendments were enacted, many observers expected the new law to usher in a futuristic system of environmental control cum economic incentives. This has yet to materialize. However, the legislation has brought in an entirely different new environmental order-rigid emissions accounting, down to each operating hour. In many respects, EPA regulation of fossil plant operations is coming more to resemble the Nuclear Regulatory Commission regulatory model for nuclear plant operations. Vol. 13011; pp. 14-18

White, J.R.
CEMs turn monitoring giant Pollution Engineering; Aug 31 1993 Description: Crucial to complying with environmental regulations is selecting appropriate pollution control equipment to capture or destroy regulated pollutants. But just as important is selecting a continuous emissions monitoring system (CEM). CEMs play a dual role in an overall compliance strategy. On one hand, they identify the type and quantity of emissions at a source as a first step for determining which regulatory requirements and control technologies are applicable.
Vol. 2513; pp. 44-46

Wilkinson, K.J.; Campbell, P.G.C.
Aluminum bioconcentration at the gill surface of juvenile Atlantic salmon in acidic media. Environmental Toxicology and Chemistry; Nov 30 1993. Description: Aluminum uptake by Atlantic salmon was examined in the laboratory at pH 4.5, under conditions similar to those found in running waters on the Canadian Precambrian Shield during spring snowmelt. Gill uptake of Al was slow, approaching steady state only after 3 d of exposure. The greatest fraction of gill-associated Al was sorbed not to the gill surface itself, but to the gill mucus.
Vol. 1211; pp. 2083-2095

Williams, A.L.; Dashek, W.V.; Vose, J.M.; Swank, W.T.
Comparison of procedures for preparation of Pinus strobus needle macromolecules. Plant Physiology, Supplement; May 31 1991. Description: Eastern white pine foliage is sensitive to adverse atmospheric O_3 and acid rain. The mechanisms by which they promote needle necrosis have not been fully elucidated. Because the literature yielded little regarding needle protein and nucleic acid contents, streptomycin sulfate (SS) and trichloroacetic acid (TCA) efficiencies for macromolecule precipitation from 1 and 2 yr-old needles of 30 yr- old trees were compared. Vol. 961; pp. 172

Williams, E.J.; Fehsenfeld, F.C.
Measurement of soil nitrogen oxide emissions at three North American ecosystems. Journal of Geophysical Research; Jan 20 1991. Description: Results from measurements of emission of nitrogen oxides from soils at three North American ecosystems are presented. These measurements were conducted during the summer and fall of 1988 at (1) a grassland site near Nunn, Colorado, (2) a coastal marine environment at North Inlet, South Carolina and (3) a deciduous forest near Oak Ridge, Tennessee. Emission of NO was highest from the grassland soil (mean: 10.0 ng N $m^{-2}s^{-1}$), intermediate at the forest area (mean: 0.28 ng N $m^{-2}s^{-1}$) and lowest at the coastal site (mean: 0.034 ng N $m^{-2}s^{-1}$). A comparison of the results from the present study with previous measurements indicates that NO_x (NO + NO_2) emission from grasslands and temperate forests are similar within each ecosystem independent of location. This suggests that simple approaches may be used to estimate soil emissions over wide areas.
Vol. 961; pp. 1033-1042

Williams, M.W.; Melack, J.M.
Precipitation chemistry in and ionic loading to an alpine basin, Sierra Nevada Water Resources Research; Jul 31 1991 Description: Wet deposition of solutes to an alpine catchment in the southern Sierra Nevada was measured from October 1984 through March 1988. Rainfall had a volume-weighted pH of 4.9, and snowfall had a volume-weighted pH of 5.3. Acetic and formic acids were important components of all wet deposition, contributing 25-30% of the measured anions in snowfall and, through analysis of charge balance deficits, the same percentage in rainfall.
Vol. 277; pp. 1563-1574

Wilson, G.V.; Jardine, P.M.; Luxmoore, R.J.; Todd, D.E.; Zelazny, L.W.; Lietzke, D.A.
Hydrogeochemical processes controlling subsurface transport from an upper subcatchment of Walker Branch watershed during storm events: Hydrologic transport processes. Journal of Hydrology; Jan 31 1991. Description: Concerns over the effects of acid rain have stimulated numerous hydrometric and geochemical studies on forested watersheds with an emphasis on stream water chemistry. However, integrated studies are seriously lacking, and inferences of soil hydrogeochemical processes from periodic stream water chemistry may be grossly misleading. A small forested subcatchment was intensively instrumented for hydrologic and chemical analyses to improve the understanding of the processes that control subsurface transport of solutes.
Vol. 123; pp. 297-316

Wirl, F.
> Pigouvian taxation of energy for flow and stock externalities and strategic, noncompetitive energy pricing. Journal of Environmental Economics and Management; Jan 31 1994. Description: The literature on energy and carbon taxes is by and large concerned about the derivation of (globally) efficient strategies. In contrast, this paper considers the dynamic interactions between cartelized energy suppliers and a consumers' government that collectively taxes energy carriers for Pigouvian motives. Two different kinds of external costs are associated with energy consumption: flow (e.g., acid rain) and stock externalities (e.g., global warming). Vol. 261; pp. 1-18

Wolcott, J.R
> Protecting the future: What Congress can do for nuclear power. NUEXCO. Monthly Report to the Nuclear Industry; Jan 31 1989. Description: Many have believed for some years that nuclear power will eventually be required in the USA despite the virtually dormant status of nuclear power construction today. It has also been recognized that important environmental questions must be confronted sooner or later. Concern with the greenhouse effect, acid rain, and ozone layer degradation have recently increased interest in a rebirth (or "reinvention" as some have termed it) of nuclear power. No. 245; pp. 25-28

Wolff, G.T.; Rodgers, W.R.; Wong, Curtis A.; Collins, D.C.; Verma, M.H.
> Spotting of automotive finishes from the interactions between dry deposition of crustal material and wet deposition of sulfate. Journal of the Air and Waste Management Association; Dec 31 1990. Description: During the summer of 1988, General Motors Research Laboratories operated a mobile atmospheric research laboratory in Jacksonville, Florida to determine the cause of environmentally-related damage that occurs on automotive finishes in many parts of the US. The damage occurs as circular, elliptical, or irregular spots that appear as deposits or precipitates. The results of the present study show that a wetting event (rain or dew) is a prerequisite for damage to occur. Vol. 4012; pp. 1638-1648

Wolfson, Z.; Spetter, H.
> Ecological aspects of east-west integration trends. Environmental Policy Review; Jan 31 1991. Description: The fall of the Berlin Wall brought about a reappraisal of the problem of integration of the territories of Eastern Europe and the USSR into the framework of the European community. In comparison with western economies the Soviet economy is said to be inefficient, lacking funds for pollution control or environment-friendly technologies and lacking in innovative capacity. As an outcome of research by the Institute of Geography of the Academy of Sciences of the USSR, a map has been prepared that locates 291 destabilized areas where populations in the USSR face grave health problems related to various kinds of pollution. Vol. 51; pp. 10-25

Woodward, D.F.; Farag, A.M.; E.E., Little; Steadman, B.; Yancik, R.
> Sensitivity of greenback cutthroat trout to acidic pH and elevated aluminum Transactions of the American Fisheries Society; Jan 31 1991. Description: The greenback cutthroat trout Oncorhynchus clarki stomias is a threatened subspecies native to the upper South Platte and Arkansas rivers between Denver and Fort Collins, Colorado, an area also susceptible to acid deposition. In laboratory studies, the authors exposed this subspecies to nominal pHs of 4.5-6.5 and to nominal aluminum concentrations of 0, 50, 100, and 300 {mu}g/L; the control was pH 6.5 treatment without Al. Vol. 1201; pp. 34-42

Xantheas, S.S.; Dunning, T.H. Jr.
> Theoretical estimate of the enthalpy of formation of HSO and the HSO-SOH isomerization energy. Journal of Physical Chemistry; Jan 07 1993. Description: The thermochemistry of HSO is of importance because this species is involved in the atmospheric oxidation of HS, one of the chemical processes that leads to acid rain. The enthalpy of formation of HSO is

estimated to be -5.4 ± 1.3 kcal/mol through a series of multireference configuration interaction (MR-CI) calculations that systematically expand the orbital basis set. Vol. 971; pp. 18-19

Yates, M.
Acid rain conference held as Congress gives final approval to bill
Public Utilities Fortnightly; Dec 06 1990
Description: This article is a report on the New Acid Rain Legislation conference held in Washington, D.C. as the final changes were being made to the Clean Air Act Amendments of 1990. The topics covered include a brief synopsis of the amendments, representative Jim Cooper's critique of the electric utility industry, EPA implementation of the amendments, emissions trading, and the preliminary results of a survey of 45 electric utilities' plans for implementing the changes.
Vol. 12612; pp. 47,49

Yates, M.
Acid rain draft study released: No environmental emergency found
Public Utilities Fortnightly; Oct 11 1990
Description: This article discusses the findings released in the draft study of the National Acid Precipitation Assessment Program (NAPAP). Some areas where acidic deposition and related air pollution is evident include 10 percent of eastern lakes and streams, visibility reduction throughout the eastern US and large metropolitan areas of the West, erosion and corrosion damage to stone and metal structures and cultural resources, and a reduction in cold tolerance of red spruce trees at high elevations. Human health effects have not been clearly demonstrated. The report tied sulfur dioxide emissions to regions of high acidic deposition. The problem is characterized as not an environmental emergency but a long term problem that should be addressed with minimum impact on other environmental goals and with efficient use of the nations economic and energy resources.
Vol. 1268; pp. 53-54

Yates, M.
Congress approves historic clean air legislation. Public Utilities Fortnightly; Dec 06 1990. Description: This article is a brief synopsis of the amendments to the Clean Air Act in 1990. Brief comments by various groups supporting and against the amendments are presented. Topics covered in the article are the trading allowance system, acid rain abatement controls, NO_x emissions reductions, clean coal technology, deletion of Wisconsin Electric Power Company provision, continuation of acid rain research programs, and EPA/NRC regulatory reponsibilities.
Vol. 12612; pp. 53-55

Yates, M.
House clean air markup begins
Public Utilities Fortnightly; Apr 12 1990
Description: This article describes the markup to the Clean Air Act. Various proposals submitted by house committee members are described covering alternative fuels, acid rain, SO_2 emissions cost sharing, and allowance trading. The clean air coalition, American Federation of Labor and Congress of Industrial Organizations and the United Mine Workers, among others, are supporting various amendments. Vol. 1258; pp. 29

Yates, M.
Senate begins clean air legislation debate
Public Utilities Fortnightly; Mar 01 1990
Description: This article reports on Senate debate on the Clean Air Act Amendments of 1989. Topics include acid rain provisions, administration objections, costs of the bill including disparity of costs in different regions and cost-sharing proposals, and the effects the current energy policy will have on the bill. Presidential, Senate, and subcommittee views on the bill are presented.
Vol. 1255; pp. 30-31, 34

Young, T.C.
Method to assess lake responsiveness to future acid inputs using recent synoptic water column chemistry
Water Resources Research; Mar 31 1991
Description: Two indices are presented for

assessing the expected responsiveness of lakes to changes in external acid inputs. The first is the ratio of the sum of acidic anion to sum of base cation equivalents (C_A/C_B), which is a familiar index of lake response to past inputs; the second quantity is buffer intensity ($\beta = -dC_A\}/dpH$), which serves as an index of lake pH sensitivity to changes in current inputs. Because the indices account for both capacity and intensity of pH buffering, the approach provides greater information on lake responsiveness to acidification than the more customary value, acid neutralizing capacity. Vol. 273; pp. 317-326

Zabinski, Catherine.; Wojtowicz, Todd.; Cole, David.
The effects of recreation disturbance on subalpine seed banks in the Rocky Mountains of Montana. Canadian Journal of Botany; May 01 2000. Description: We investigated the soil seed bank in a subalpine ecosystem with patchy disturbance from camping. Soil cores were collected from three site types, heavily impacted, lightly impacted, and undisturbed, that differed in area of bare ground and depth of surface organic matter. We hypothesized that the density and composition of the seed bank would vary with depth of surface organic matter and distance from established vegetation. Vol. 78; no. 5; pp. 577-582

Zhang, Jiazhong; Millero, F.J.
The rate of sulfite oxidation in seawater Geochimica et Cosmochimica Acta; Mar 31 1991. Description: The oxidation of sulfite in the atmosphere has long been a subject of interest because it is important in the formation of acid rain. Unlike the most volatile sulfur compounds, such as dimethyl sulfide, the direction of the net flux for SO_2 is from the atmosphere into the oceans. Vol. 553; pp. 677-685

Zhang, W; Jia, M; Yu, J; Wu, T.; Yahiro, Hidenori; Iwamoto, Masakazu
Adsorption properties of nitrogen monoxide on silver ion-exchanged zeolites Chemistry of Materials; Apr 30 1999 Description: The removal of nitrogen oxides (NO_x) which cause acid rain and air pollution is an important global environmental problem which needs to be solved soon. The adsorption properties of nitrogen monoxide (NO) on various silver ion-exchanged zeolites were examined by adsorption-desorption measurements in a fixed bed flow apparatus. Both reversible (q_{rev}) and irreversible (q_{irr}) adsorption of NO is dependent on the aluminum content in the zeolites and the zeolite structure. The amounts of reversible and irreversible adsorption of NO per silver ion increased with decreasing aluminum content of the zeolites and were constant, independent of the ion exchange level for ZSM-5 zeolites, but increased with the ion exchange level for mordenite zeolites (MOR). Vol. 11; no. 4; pp. 920-923

Zilberman, D.
Environmental aspects of economic relations between nations. American Journal of Agricultural Economics; Dec 31 1992. Description: The last 30 years have seen growing concern about environmental issues and a continuous introduction of new environmental regulations. At present, global environmental issues, such as global warning, acid rain, and the disappearance of tropical rain forests, have become paramount. Furthermore, the impact of environmental regulations on international trade patterns is a related area of concern, with many analysts believing that countries may try to use environmental and safety regulations as barriers to free trade Vol. 745; pp. 1144-1149

Zimmer, M.J.
Clean air and project financing. Independent Energy; Jan 31 1992. Description: This article examines how environmental requirements are challenging the developers ability to secure financing for independent energy projects. The topics addressed in the article include a review of the US Environmental Protection Agency auction rules for acid rain emission allowances, short term and long term market demand, project financing issues, credit value and matching interests. Vol. 221; pp. 21-24

Zink, J.C
    Burner swirls NO$_x$ away. Power Engineering (Barrington); Nov 30 1997.
    Description: Cleaner boilers that will help reduce acid rain, photochemical smog and tropospheric ozone are now coming on-line, partly as a result of a new burner design for power-generation boilers based on MIT research and now commercialized under exclusive license to ABB C-E Services Inc. The rapidly stratified flame core (RSFC) burner achieved very large nitrogen oxide reductions of up to 90% with natural gas as fuel, and 70 to 80% when burning pulverized coal and heavy fuel oil, respectively. The RSFC design provides a fuel-rich (oxygen deficient), high-temperature environment early in the flame to allow the chemical conversion of the NO$_x$ precursors to harmless molecular nitrogen. A subsequent lower-temperature, fuel-lean environment, in which the remainder of the air is mixed with the remaining fuel, then ensures complete combustion. Vol. 101; no. 12; pp. 47-56

Zmuda, J.T.
    Acid rain compliance planning: Compliance issues and options Proceedings of the American Power Conference; Jan 31 1992.
    Description: The difficulty of establishing a compliance strategy was demonstrated at a recent technical seminar on the Impacts of the Clean Air Act conducted by the R-C Institute. Attendees were grouped into teams and assigned the task of developing a simplified compliance strategy for a hypothetical utility. Teams were composed of representatives form the utility sector, state regulators, and engineering firms. The strategies developed by each team were presented for review by the seminar participants. This paper will discuss the various compliance strategies developed by the teams. As a preface to the Case Study, the authors will briefly examine several of the issues which must be addressed to develop these strategies.
    Vol. 541; pp. 371-376

AUTHOR INDEX

A

Abate, T, 37
Abelson, P.H., 37
Achi, C.A., 117
Adams, M.B., 37
Ahmad, S., 96
Albers, P.H., 38
Albritton, D.L., 38
Alewell, C, 38
Alfrey, A.C., 67
Alkezweeny, A.J., 38
Allen, H.L., 106
Allen, J.M., 66
Allen, Robert.B., 39, 56
Allison, J.E., 39
Alm, L.R., 39
Alvarez-Hernandez, X., 105
Amokrane, H., 39
Anastasio, C., 66
Andersen, S., 114
Anderson, P.D., 40
Andersson, T., 75
Annamalai, S, 41
Aoki, Ichiz, 40
Aono, Hiromichi, 40
Appanna, V.D, 40
Appleton, E.L, 40
Arenillas, A, 129
Aris, R., 40
Armstrong, A., 133
Armstrong, A.Q., 64
Arpin, Pierre, 122
Asai, Ei-Ichiro, 41
Ashmore, M.R., 100
Asolekar, S.R., 41
Atherton, C.S., 120
Ayers, G.P., 41

B

Baedecker, P.A., 41
Baker, R., 93
Bandyopadhyay, J.K, 41
Banta, H.M., 41
Banwart, W.L., 42
Barbosa, M.A., 60
Barker, A.V., 82
Barnard, W., 71
Barnes, M.W., 45
Bartels, C.W., 42
Barth, M.C., 42
Bartlett, R.J., 58
Basman, A.R., 42
Bassett, S.M, 43
Bates, T.S., 43
Batterman, S., 43
Baublis, D.C., 43
Bauer, F.W., 43
Baxter, James.W., 44
Baylor, J.S., 44
Beamon, J.A., 44
Begley, R., 44
Behnke, Robert.J., 119
Bélanger, Luc, 44
Bellehumeur, C., 44
Bencala, K.E., 45
Benner, R., 133
Bennett, R.R., 45
Berg, Björn., 147
Berger, Torsten.W., 45
Bernow, S., 45
Berrang, P, 49
Bérubé, Pierre, 96
Bevington, S.R., 131
Bhatti, N., 46
Blake, M.W., 64

Blanchar, R.W., 57
Blasio, C, 125
Blew, R.D, 46
Boatman, J.F., 46
Boettcher, J, 47
Bohonak, A.J., 47
Bolton, Kim A., 95
Bourdon, J.C., 47
Boutacoff, D., 47
Bowersox, V.C., 48, 103
Bowes, S.M. III, 48
Brandt, C, 48
Brauer, M., 139
Bredemeier, M., 38, 99
Breemen, Nico.van, 48
Britton, K.O., 49
Brook, J.R, 49
Brooks, R.T., 49
Brotons, Lluís, 49
Brown, A., 111
Brown, G.W., 50
Brown, S.D., 77, 78
Brumbaugh, W.G., 61
Brune, W.H., 50
Bube, R.H., 50
Buhr, M.P., 50
Bukaveckas, Paul., 50
Bulger, Arthur J., 51
Burger, J., 51
Burkhardt, D.A., 51
Burkhart, L.A., 51
Burris, L.H. Jr., 64
Burton, A.J., 52, 104
Buttorff, L.A., 116

C

Cahill, T.A., 52

Camiré, Claude, 83
Campbell, C.A., 59
Campbell, D.H., 52
Cannon, W.N. Jr., 52
Carignan, Richard., 53
Carreiro, Margaret.M., 44
Carson, C., 115
Cascio, T.B., 53
Cason, T.N., 53
Castelle, A.J., 53
Caton, J.E., 53
Chandler, A.S., 55
Chandra, Peeyush, 70
Chang, J.S., 122, 141
Chapman, E.G., 54
Charlebois, D., 117
Chaumerliac, N., 85
Chen, L., 148
Chen, Y.P., 119
Cheplick, G.P., 54
Chesnoy, A.B., 54
Chevone, B.I., 97
Chorley, G.B., 42
Choularton, T.W., 55
Chowdhury, B.H., 54
Christian, D., 40
Christophersen, N., 114
Church, M.R., 80
Clair, T.A., 54
Clark, E., 55
Clark, J.M., 107
Clark, P.A., 55
Clegg, S.L., 108
Clinton, P.W., 56
Clinton, Peter.W., 39
Cluis, D.A., 106
Coichev, N., 56
Colbourn, E.L., 119
Cole, F., 56
Cole, M.A., 111
Coleman, W.G., 56
Collins, D.C., 151
Collins, D.J., 57
Collins, M., 57
Collins, P.V., 54, 57
Conkling, B.L., 57
Connick, R.E., 57
Cooper, K., 51
Corcoran, E., 57
Cosby, Bernard.J., 51
Costella, M.P., 83
Cozza, A., 58
Cronan, C.S., 58

Cunjak, R.A., 59
Curtin, D., 59
Czarnecki, J., 59

D

Dahlgren, R.A., 59
Dahlin, R.S., 59
Dai, Q., 60
Dale, V.H., 146
Dalledone, E., 60
Darveau, Marcel, 60
Dashek, W.V., 150
David, M.B., 60, 93
Davidovits, P., 88
Davis, D.D., 61
Davis, P.N., 61
de Steiguer, J.E., 87
Deckert, Ron.J., 61
Dehayes, D., 61
DeLonay, A.J., 61
Desai, M.S., 62
DesGranges, Jean-Luc, 60
Devitt, T.W., 62
Dignon, J., 120
Dixit, S.S., 90
Djuric, M., 62
Dobson, J.E., 62
Dougherty, W., 45
Dragovich, D., 63
Drdla, E., 63
Dreschel, T.W., 63
Drever, J.I., 118
Drexhage, Michael., 63
Driscoll, C.T., 58, 63, 107, 132
Duan, L., 77
Duchesne, L.C., 110
Dudek, D.J., 63, 64

E

Ecsedy, C., 116
Edwards, N.T., 110
Edwards, R.A., 64
Egyed, M., 126
Ehara, Yoshiyasu, 117
Ek, A.S., 64
Elless, M.P., 64
Ellis, H., 65
Ember, L.R., 65
Evans, A. Jr., 65

F

Fahrer, S., 65
Farag, A.M., 61, 66, 151
Fasth, W.Jc
Faust, B.C., 66
Fehsenfeld, F.C., 38, 50
Feldman, P.L., 95
Fendick, E.A., 131
Fenn, M.E, 66
Fernandez, I.J., 60
Ferrús, Lluís, 49
Fey, M.V., 66
Finke, R.L., 42
Finzi, Adrien C., 48
Fiss, F.C., 66
Fitzgerald, W.F., 67
Fjeld, Eirik, 137
Flanigan, T., 67
Flaten, T.P., 67
Fleyfel, F., 67
Fornes, R.E., 149
Forti, M.C., 67
Fournier, R.E., 68
Frank, R., 48
Freedman, A., 60
Freedman, B., 68, 110
Friedmann, A.S, 68
Frigione, G., 135
Frisbie, M.P., 68
Fu, Ji-Meng, 69
Fuente, E, 129
Fuller, R.D., 60

G

Gagen, C.J., 69
Galloway, J.N., 69
Gardner, J.A., 88
Garten, C.T. Jr., 69
Gatz, D., 71
Gauri, K.L., 70
Geissler, M., 70
Gephard, Stephen.R., 119
Gervat, G.P., 55
Ghan, S.J., 120
Ghosh, Mini, 70
Ghuman, G.S., 70
Giamello, E., 71
Giesler, R, 71
Gilbert, R.D., 149
Gillette, D.A., 71
Gilliam, Frank S., 71

Gjessing, E.T., 72
Goldstein, B.D., 72
Gordon, D., 72
Gough, William A., 95
Graveland, Jaap, 72
Grebenyuk, O.V., 72
Grebenyuk, V.D., 72
Greenberger, L.S., 73
Griffith, M.B., 73
Grodzin´ska-Jurczak, M., 73
Groffman, P. M., 73
Grupenhoff, J.T., 73
Guinee, J.B, 74
Gunn, John.M., 94
Gunter, L., 46
Günthardt-Goerg,
   Madeleine.S., 74
Gunther, A.J., 74
Gupta, G., 74
Gurbin, G.M, 75
Gureev, A.A., 75
Guruswamy, L.D, 75

H

Haan, H. De, 136
Hahn, R.W., 75
Håkanson, Lars, 75, 76
Halkos, G.E., 76
Hallett, R.A., 76
Hamburg, Steven.P., 99
Handley, C.O. Jr., 76
Hannapel, J.S, 92
Hansen, B.K, 76
Hansen, D.A., 77
Hanson, B.M., 77
Hao, J.M., 77
Harding, A.W., 77, 78
Hargeby, A., 78, 94
Hart, G.S., 78
Hassett, J.J, 136
Hata, Kunihiko, 41
Hathaway, A.M. II, 78
Hauhs, M., 114
Hedin, L.O, 79
Heiderscheit, J., 79
Hemond, H.F., 79
Hendershot, W.H., 79
Henrichs, R., 80
Herlihy, A.T., 80, 90
Heslin, J.S., 80
Hessen, D.O., 80
Hicks, B., 81

Hill, M.W., 107
Hill, T.A., 55
Hille-Salgueiro, M., 109
Hinkle, C.R., 63
Ho, C.H., 53
Hobbs, B.F., 81
Hoelldampf, B., 82
Hofmann, D.J., 82
Hoisve, R.A., 82
Holdren, G.R. Jr., 81
Holtze, K.E., 90
Holzman, D., 82
Honious, J.C., 81
Hordijk, L., 83
Hoske, M.T., 83
Hough, A.M., 83
Houle, Daniel, 83
Houpis, J.L.J., 83, 115, 116
Hov, Oe, 84
Hsia, Yue-Joe, 99
Hu, W., 96, 115
Huang, J., 40
Huang, Wenxiong, 84
Huang, Y.J., 84
Hubbard, H.M., 84
Hudson, J.G., 84
Hughes, R.N., 85
Huret, N., 85
Hutchinson, Thomas C., 68, 85

I

Igolkina, E.D., 86
Irwin, B., 86
Isaka, H., 85
Ishihara, Shigehis, 86
Ishizuka, T., 86
Itoh, Mik, 86
Iwamoto, Masakazu, 87

J

Jager, H.I., 87
Jardine, P.M., 150
Jayanty, R.K.M., 87
Jayne, J.T., 88
Jeffries, D.S., 88
Jia, M, 153
Johansson, C., 127
Johansson, Per, 88
Johnson, B.D., 88
Johnson, Chris.E., 88

Johnson, D.W., 89
Johnson, G.L., 89
Joskow, P.L., 89
Joslin, J.D., 89

K

Kabashima, H., 86
Kaiser, J, 89
Kaitala, V., 144
Karagatzides, Jim D., 95
Kaufman(n), P.R., 80, 90
Kaufman, Y.J., 90
Keller, W., 90
Kelly, C, 96
Kelly, J.M., 58, 89
Kern, E., 91
Kieber, R.J, 91
Kim, A., 91
Kim, D.S., 91
Kim, Y.J., 91
King, D.J., 62
King, G.A., 92
King, Hen-Biau, 99
Kirkwood, D.E., 92
Kishida, Haruo, 117
Kissam, A.D., 92
Kline, T.R, 92
Knotkova, D, 92
Kohut, R.J., 96
Kong, F.X, 92
Konovalova, I.D., 72
Koop, T, 110
Kortelainen, P., 93
Kosuri, S.R., 144
Kouterick, K.B., 93
Kovacik, J.M., 93
Kowalok, M.E., 93
Kramer, J.R., 57
Kress, M.W., 93
Kreutzweiser, David.P., 94
Krouse, H.R., 140
Kruger, J, 94
Krupa, Sagar.V., 94
Kullberg, A., 78, 94
Kulmala, M., 94
Kulshrestha, U.C., 95
Kumar, Ashij J., 95
Kumar, K.S., 95
Kumar, N., 95
Kumar, S., 95
Kumari, K.M., 95
Kuperman, R.G., 142

Kutka, F.J., 95
Kut'kov, V.S., 129
Kwong, K.V., 96

L

Lachance, Stephanie, 96
Lam, D.CL., 88
Lambert, D., 96
Latona, M., 96, 115
Lauer, S, 47
Laulainen, N., 46
Laurence, J.A., 96
Lave, L., 97, 129
Lawson, D.R., 97
Leaf, D.A., 97
Lee, B., 97
Lee, R., 46
Lee, Shaoyung, 57
Lee, W.S., 97
Leiter, J.C., 68
LeMay, B., 98
Leopold, D.J., 104
Levin, M.H., 98
Levine, J.S., 99
Leygraf, C., 144
Liechty, H.O., 104
Lin, Teng-Chiu, 99
Lindberg, S.E., 99
Linden, H.R., 99
Ling, K.A., 100
Little, E.E., 61, 66, 151
Liu, B.J., 77
Lobsenz, G., 100, 101
Lock, R., 101
Loeb, A.P., 102
Loerting, Thomas, 102
Long, J., 102
Long, R.Q., 102
Loucks, O.L., 102
Lovett, G.M., 102
Lovins, K, 125
Lucas, A, 103
Luecken, D., 46
Luley, C.J, 103
Lux, Heidi B., 103
Luxmoore, R.J., 150
Lydersen, E., 103
Lynch, J.A., 103
Lynch, T.R., 123
Lynham, T.J., 110

M

MacDonald, N.W., 104
Mackun, I.R., 104
MacLeod, C., 104
MacQuarrie, D.M., 104
Madigosky, S.R., 105
Madsen, B.C., 63, 105
Magnacca, G., 71
Magrans, Marc, 49
Maibodi, M., 105
Malcolm, R.L., 105, 106
Maljanen, M., 125
Mallory, M.L., 106
Mänd, Raivo, 106
Marcotte, D., 44
Markewitz, D, 106, 107
Markey, E.J., 107
Marsh, A.R.W., 55
Marteinz, O, 129
Martel, Jocelyn., 60
Martin, C.W., 107
Martin, L.R., 107
Martin, P.J.S., 107
Martin, R.B., 107
Massey, D.D, 108
Massucci, M., 108
Matzner, E., 38, 108
Maull, L.A., 63
Mavinic, D.S., 104
McAvoy, D.C., 59
McCally, M., 108
McClean, Therese.M., 111
McCormick, L.H., 108
McCormick, Stephen.D., 119
McCormick, W.T. Jr., 109
McDonald, A., 45, 109
McHenry, J.N., 109
McIlvaine, R.W., 109
McIntosh, H., 109
McKenna, M.A., 109
McKinney, K., 111
McLaughlin, J.W., 110
McLaughlin, S.B., 110
McMillen, R., 81
McNicol, D.K., 106
McQuattie, Carolyn J., 74
McRae, D.J., 110
Meilinger, S.K, 110
Meissner, R.E. III, 96
Menz, F.C, 110
Menzel, R.G., 111
Messer, J.J., 90

Middleton, P., 141
Miller, K.W., 111
Miller, R.W, 111
Miller, Steven.L., 111
Mine, Yosihiro, 87
Mitch, M.E., 90
Mitchell, M.J., 60, 140
Mitnick, S., 111
Mitusova, T.N., 75
Mizuno, Noritaka, 87
Molau, Ulf., 111
Molburg, J.C., 112
Moldan, F, 71
Mookerjee, G., 112
Moore, A., 112
Moore, T, 112
Morales-Baquero, Rafael, 112
Morgan, M., 113
Mori, Yoshiaki, 40
Morrison, I.K., 68
Morterra, C., 71
Moyad, A., 113
Moynihan, D.P., 113
Mudur, G, 113
Muenster, U., 114
Mulder, J., 114
Muller, P.K., 117
Munn, Ted, 114
Murphy, D., 71
Myers, N., 114
Mykkelbost, T.C., 114

N

Nagelhout, M., 115
Nan, G.D., 115
Nelson, G.L., 115
Neufeld, H., 121
Neufeld, R.D., 96, 115
Neuman, L., 115, 116
Newton, R.M., 58
Ng, W.S., 116
Nikolaidis, N.P., 116, 117
Nishihara, Hirosh, 86
Nomura, Toshikazu, 71, 117
Nordhaus, R.R., 117
Norton, R.B., 50
Norton, Stephen.A., 137
N'soukpoe-Kossi, C.N., 117
Nykänen, H., 125

O

Odnevall, I, 117
Olem, H., 116, 118
Omorjan, R., 62
Ondrey, G., 118
Orchard, D., 118
Ouimet, Rock, 83
Oxburgh, R., 118

P

Padmanabha, A.P., 118
Palmer, G.W.R. Si, 75
Palola, E., 118
Paquin, Raynald, 83
Parkinson, G., 118
Parrish, D.D., 50
Parrish, Donna.L., 119
Patel, A., 119
Patrizia, C.A., 119
Patt, A., 119
Penner, J.E., 120
Pennypacker, S.P., 93
Pepper, G.E, 136
Pereira, C.J., 59
Perry, S.A., 73
Peterman, R.M., 120
Petersen, R.C. Jr., 78, 94
Peterson, C.E. Jr., 120
Phillips, K., 121
Phillips, R.A., 121
Phillips, V.D., 121
Pickett, Steward.T., 44
Pier, P.A., 121
Pis, J.J., 129
Pleim, J.E., 122
Plummer, P.L.M., 122
Polasek, S., 98
Poleo, A.B.S., 103
Pollard, Heather.G., 94
Pollock, T.L., 54
Ponge, Jean-François., 122
Pool, R., 122
Popova, T.V., 123
Popp, C.J., 123
Porter, J.M, 92
Porter, P.M., 42
Powell, B.E., 123
Power, M., 123
Pregitzer, K.S., 52, 104
Prenzel, J., 123, 124
Presentación Carrillo, 112
Prowse, T.D., 59
Punuru, A.R., 70
Pye, J.M., 87

Q

Quillian, A.M., 124

R

Rabl, V.A., 124
Rafizadeh, H.A., 124
Rahman, S, 124
Rajaretnam, G, 125
Ramsay, S. L., 125
Ranogajec, J., 62
Reams, G.A., 118
Reche, Isabel, 112
Reddy, M.M., 134
Reed, Austin, 44
Reed, D.D., 110
Rees, T.H., 125
Regina, K., 125
Renner, R, 125
Reynolds, W.E., 126
Rhiner, Claudia., 74
Rhines, M.F, 91
Rhorer, J.R. Jr, 126
Richardson, C.J., 131
Richardson, G.M., 126
Richardson, H.H., 67
Richter, D.D., 106, 107
Rickett, B.I, 126
Rinallo, C., 127
Rittenhouse, R.C., 127
Roderick, D., 50
Rodgers, L.M., 127
Rodgers, W.R., 151
Rogt, R.D., 114
Roila, T., 93
Romanowicz, Rachel.B., 88
Rondon, A., 127
Rooth, R.A., 127
Rossin, A.D., 128
Rostam-Abadi, M., 128
Rostorfer, C.R., 128
Roy, K.A., 128
Rozanov, I.A., 129
Rubiera, F, 129
Rubin, E.S., 129
Ruch, R.B. Jr., 130
Rush, R.M., 62
Russell, M., 130
Ryan, B.D., 130

S

Saboni, A., 39
Sadler, B., 130
Sager, Eric.P., 85
Salbu, B., 103
Sale, M.J., 87
Samson, P.J, 49, 130
Sandell, M., 131
Sarosdy, R.L., 131
Sase, Hiroyuki, 131, 144
Sasek, T.W., 131
Saunders, G.L., 132
Savastano, C.A., 132
Saxena, A., 95
Saylor, R.D., 132
Schaefer, D.A., 99, 132
Schakenbach, J.T., 132
Scheidegger, Christoph., 74
Scherzer, A.J., 133
Schindler, D.W., 133
Schlegel, D.M., 133
Schnoor, J.L., 117
Schoenberg, S.A., 133
Schofield, C.L., 58
Schrader, E.L., 134
Schuette, R., 66
Schuh, S.A., 134
Schuster, P.F., 134
Segal, M., 134
Seiler, J.R., 121
Seip, H.M., 146
Selin, Henrik., 135
Seltzer, R., 135
Sersale, R., 135
Shah, A.Y, 135
Sharp, R.G., 136
Sharpe, W.E., 69, 136
Shaw, P.J., 136
Shen, S., 136
Sheppard, D., 40
Shih, H.C., 147
Shioya, Y., 71
Shortle, W.C., 136
Showman, R.E., 137
Shuping Bi, 46
Sievers, R.E., 50
Sigurd, Rognerud, 137
Silvola, J., 125
Sinclair, P.C., 71
Singh, N., 137

Singh, S.N., 137
Sisterson, D.L., 137
Sisterson, D.L., 48
Sjoestroem, J., 138
Skelly, J.M., 61, 93
Sklar, S., 138
Skotvold, Trond, 137
Smith, K.T., 136
Smith, L., 138
Smith, R.B., 138
Snyder, T.R., 59
Sobecky, P., 133
Sobolevskaya, T.T., 72
Sokolov, V., 138
Sokolov, V.V., 75
Solomon, B.D., 138, 139
Sondag, Francis, 122
Sorini, S.S., 139
Spainhoward, D.A, 139
Spengler, J.D., 139
Spiegel, R.J., 140
Spirkina, N.P., 75
Srivastava, R.K., 140
Stalhandske, P., 78
Stam, A.C., 60, 140
Stanton, Nancy L., 111
Stauffer, R.E., 140, 141
Steadman, B., 66, 151
Steele, K.F., 147
Stein, H., 141
Stensland, G.J., 71
Stevens, P.S., 50
Stigliani, W.M., 141
Stockwell, W.R., 141
Strandberg, H, 142
Strebel, O, 47
Streets, D.G., 46
Stricker, G.D., 142
Stubbs, C.S., 142
Sugg, P.M, 142
Sugimoto, Eisuke, 40
Suomela, Janne, 143
Susfalk, R.B., 89
Svensson, J.E, 143
Svensson, M., 78
Sverdrup, H.U., 79

T

Tabors, R.D., 143
Tahvonen, O., 144
Tai, A.F., 107
Tajchman, S.J., 144

Takamatsu, T., 144
Takamatsu, Takejiro, 131
Tanabe, K., 86
Tanda, Kenji, 87
Thompson, Dean.G., 94
Thorneloe, S.A., 140
Thornton, F.C., 121
Tidblad, J., 144, 145
Tilgar, Vallo., 106
Timmerman, Peter, 114
Todd, D.E., 150
Todd, D.M., 145
Tombach, I., 145
Tonnessen, K.A., 145
Traina, J., 145
Trocciola, J.C., 140
Trottier, C., 117
Tsujita, Cameron J., 146
Turk, J.T., 52
Turner, R.S., 81, 146

U

Uberoi, M., 59

V

Vaghjiani, G.L., 146
Valentine, R.L., 41
Valin, V., 46
Vallejo, L.E., 115
Van Deusen, P.C., 118
Van Dreason, R., 132
Vance, G.F., 60
Veretennikova, T.N., 75
Verhage, A.J.L., 127
Vishnyakova, T.P., 123
Vlckova, J, 92
Vogt, R.D., 114, 146
Vong, R.J., 146
Vose, J.M., 150

W

Wagman, J.D., 147
Wagner, G.H., 147
Walker, R.F., 147
Walse, Charlotta, 147
Walton, J.J., 120
Wang, J.H., 147
Wang, Lih-Jih, 99
Wang, S.X., 77
Wang, X., 148

Warfvinge, P., 79
Warkentin, D, 148
Watmough, Shaun A., 85, 148
Webb, J.R., 80
Weinstein, D., 148
Weller, R., 148
Weng, Chengyu., 149
Westenbarger, D., 149
White, D.F., 149
White, J., 111, 149
White, J.R., 149
Wigington, P.J. Jr., 80
Wilkinson, K.J., 150
Willey, J.D, 91
Williams, A.L., 71, 150
Williams, E.J., 150
Williams, M.W., 150
Williams, R.T., 53
Willyard, B., 98
Wilson, G.V., 150
Wirl, F., 151
Witter, J.A., 104
Wojtowicz, Todd., 153
Wolcott, J.R, 151
Wolff, G.T., 151
Wolfson, Z., 151
Wong, Curtis A., 151
Wong, I., 88
Woodward, D.F., 66, 151
Woodward, F., 61
Worsnop, D.R., 88
Wu, T., 153

X

Xantheas, S.S., 151

Y

Yahiro, Hidenori, 87, 153
Yamaguchi, T., 86
Yan, N.D., 90
Yates, M., 152
Yoshida, Tomio, 131
Young, T.C., 152
Yu, J, 153
Yuegang Zuo, Hoigne J., 87
Yunus, M., 137
Yurechko, V.V., 123
Yurish, Bradley.M., 71

Z

Zabinski, Catherine, 153
Zahniser, M.S., 88
Zelazny, L.W., 150
Zellweger, G.W., 45
Zhang, Jiazhong, 153
Zhang, W., 153
Zhou, X., 77
Ziegler, E.L., 42
Zilberman, D., 153
Zimmer, M.J., 153
Zink, J.C., 154
Zmuda, J.T., 154
Zobens, V., 68

SUBJECT INDEX

A

absolute principal component analysis (APCA), 69
acetic acid, 96
acid deposition, vii, xi, xiii, xv, xviii, xix, 1, 8, 10-12, 19, 20, 24, 28, 29, 33, 34, 38, 44, 49, 51, 52, 66, 74, 84, 87, 94, 100, 109, 115, 117, 120-126, 129, 132, 141, 142, 146, 151
acid neutralizing capacity (ANC), 19, 20, 47, 58, 60, 79, 80, 87, 90, 106, 132, 143, 153
acid precipitation(s), x, xii, xviii, 39, 55, 71, 72, 76, 77, 79, 80, 82, 83, 109, 114, 116, 118, 135, 136
Acid Rain Program, 6, 9, 12-15, 17-19, 21, 22, 53, 94, 120, 125, 133, 138
acidic aquatic ecosystems, 6
acidic compounds, 2, 4, 8
acidic deposition, x, xiii, xviii, xix, 4, 5, 6, 8, 31, 33, 37, 41, 49, 53, 58, 59, 80, 81, 88, 89, 93, 102, 103, 104, 116, 120, 126, 133, 136, 143, 144, 145, 152
acidic lakes, 4, 5, 11
acidic streams, 4, 5
acidic substances, 3, 31
acidic water(s), vii, 1, 5, 7, 79, 117, 132
acidification, x, xii-xiv, xvi-xix, 3-6, 10, 14, 20, 21, 24, 28, 35, 38, 47, 50, 51, 54, 59, 61, 62, 64-67, 71, 72, 74, 78, 80, 85, 87, 89, 90, 93, 106, 107, 109, 113, 116, 118, 123, 124, 132, 136, 141, 143, 146, 147, 153
Adirondack Mountains, 4, 5, 6, 62
air pollution, xv, 8, 10-12, 29-34, 40, 48, 53, 56, 60, 72, 73, 78, 87, 91, 96, 102-104, 112, 117, 124, 127-129, 134, 139, 140, 145, 152, 153
algae, 6, 21, 93, 133
alkalinity, ix, xii, 58, 74, 79, 82, 105, 117, 121, 123

allowance allocation, 16, 26
allowance system, 13, 23, 24, 26, 27, 53, 105, 152
Allowance Tracking System (ATS), 15, 79
allowance transactions, 15, 17, 100
alternative emission limitation (AEL), 17
aluminum levels, 5
aluminum, ix, x, xii, xv-xx, 4, 5, 7, 21, 31, 35, 58, 60, 65-69, 92, 103, 105, 108, 115, 136, 151, 153
ambient sulfate, 109
anaerobic digester gas (ADG), xiv, 140
anthropogenic factors, 29, 32
Appalachians, xv, 4, 33, 61, 80, 89
artificial neural network (ANN), 41
atmospheric boundary layers (ABL), 135
atmospheric deposition gradient, 108
atmospheric deposition, ix, xi, xvii, xviii, xx, 6, 40, 49, 52, 63, 69, 80, 81, 89, 90, 96, 99, 102, 108, 110, 113, 116, 121, 125, 137
autoclaved cellular concrete (ACC), 96, 115

B

battery acid, 3
biodiversity, 56, 75, 142
Blue Ridge Mountains, 30, 87
branch exposure chambers (BECs), 116
buffering capacity, vii, 1, 4, 7, 20, 59, 123, 147
bulk precipitation, 45, 46, 83, 108
Bush, President George, 39, 130, 136, 138

C

cadmium, 31
carbon bonded sulfur (CBS), 113
carbon compounds, 31
carbon dioxide, xvi, 2, 10, 128, 147

Catskill Mountains, 4, 7
charcoal filtered (CF), 116
charcoal-filtered air (CFA), 38, 52
chemical constituents, 20, 38, 137
chemical plants, 52
chemical process industries (CPI), xvi, 57, 96
chemical reactions, 21, 31, 50, 120
chronic acidification, 6
Clean Air Act (CAA), 9, 13-15, 18, 28, 30, 34, 37, 42-44, 46, 51-53, 56-62, 64-66, 79, 81, 83, 84, 86, 96, 97, 98, 101, 102, 105, 107, 109, 111, 112, 115, 117-119, 125-133, 136, 138, 139, 148, 149, 152, 154
Clean Air act Amendments (CAAA), 16, 23, 24, 51, 57, 59, 66, 83, 115, 119, 125, 126, 131, 136, 138, 139
Clean Air Status and Trends Network (CASTNET), 2
cloud condensation nuclei (CCN), 85, 90
coal combustion, 78, 91
coal industry, 108
coal prices, 108, 126
coal-fired utility boilers, 13
coarse woody debris (CWD), 56
coastal ecosystems, 6
combustion of coal, 77, 78, 129
combustion process, 77
compliance methods, 83, 139
compliance requirements, 112
compliance strategy, 73, 100, 149, 154
Comprehensive Sulfate Tracking Model (COMSTM), 109
congressional debate, 30, 34
continuous emission(s) monitoring (systems) (CEM(S)), 13, 17, 112, 126, 127, 133, 149
corrosion inhibitors, 105

D

Data Acquisition System (DAS), 112
deforestation, 53, 119, 121, 134
Department of Energy (DOE), 26, 28, 37, 55, 56, 64, 113, 138
diseases, 29, 32
dissolved inorganic carbon (DIC), 81
dissolved inorganic nitrogen (DIN), 112
Dissolved organic carbon (DOC), 80, 93, 94, 106, 114
domestic politics, xiv, 39
dry deposition, vii, xiv, 1, 2, 8, 20, 32, 45, 69, 77, 81, 83, 124, 134, 144, 151

E

Eastern Europe, 78, 135, 151
ecological changes, 6, 21
ecological effects, 4, 146
ecosystem alkalinity, 69
Electric Power Research Institute (EPRI), 47, 54, 77, 101, 143
electric utilities, 44, 54, 58, 62, 67, 81, 84, 112, 115, 117, 119, 131, 133, 138, 139, 141, 152
Electricity Consumer's Power to Choose Act, 45
emission allowances, xiii, xviii, 51, 55, 61, 101, 115, 138, 139, 153
emission limitations, 16, 17, 27, 44, 62
Emissions Tracking System, 17
energy balance, 69
energy efficiency, xviii, xix, 12, 16, 67, 109, 144, 148
Energy Information Administration (EIA), 44, 100
environmental benefits, 13, 14, 45, 94
environmental damage, 134
Environmental Defense Fund (EDF), xii, 40, 101
environmental fallout, 8
environmental policy, xiii, 23, 44, 63, 73, 100
environmental pollution, 86
environmental problems, 47, 56, 73, 77, 83, 86, 131, 134, 136
Environmental Protection Agency (EPA), 1, 2, 10-12, 15-19, 22, 24, 26-29, 33, 37, 40, 43, 44, 51-53, 58, 65, 66, 87, 91, 94, 97, 100-103, 107, 109, 112, 120, 125-127, 131, 136, 138-140, 144-146, 149, 152, 153
environmental protection programs, 19
environmental stressors, 6
episodic acidification, 4, 132
ethylene, 31
eutrophication, 6, 21, 76
excess emissions, 17, 27
exchangeable base, 89

F

Federal Energy Regulatory Commission (FERC), 102, 117, 138
fertilization, 42, 59, 66, 122
fish, vii, 1, 3, 5, 6, 11, 14, 19, 21, 48, 51, 61, 66, 67, 69, 70, 72, 74, 75, 82, 96, 126
fixed price, 15, 27
flue gas desulfurization (FGD), 57, 86, 98, 105, 116, 129, 134, 135, 140, 145
fluoride, 31
forest decline, x, xix, 30, 32, 60, 63, 85, 120

forest declines, 32
forest health and productivity, 33
Forest Service, 29, 32, 33, 52, 109, 116
forests, x, xii, xiv, xv, xvii, 3, 4, 6, 7, 11, 14, 29-35, 37, 48, 52, 66, 76, 79, 82, 85, 87-89, 99, 107, 108, 110, 115, 121, 127, 143, 146, 150, 153
forward-scattering spectrometer probe (FSSP), 91
fossil fuels, 1, 8, 10, 11, 21, 22, 31, 47, 54, 56, 58, 68, 90, 91, 102, 139, 147
fresh weight increment (FWT), 98
freshly painted vehicles, 8
fuel cell (FC), 140
fuel consumption, 14, 25, 27
furnace sorbent injection (FSI), 116

G

gas deposition, 20
gaseous pollutants, 120, 145
gasoline, x, 11, 21, 31, 72
geographic information systems (GIS), 62
Global Tropospheric Experiment (GTE), 67
groundwater, 47, 65, 71, 72, 76, 111, 138
growth processes, 35
growth reductions, 30, 31, 33, 87

H

harmless gases, 86
hazardous air pollutant (HAP), 52
heat recovery steam generator (HRSG), 93
human health, 3, 9, 10, 11, 58, 64, 72
Humic Lake Acidification Experiment (HUMEX), 54, 57, 72, 78, 80, 93, 94, 113, 114, 136, 146
hydrogen, xiii, xvi, xix, 20, 41, 42, 47, 62, 63, 66, 72, 75, 86, 87, 91, 95, 96, 103, 109, 125, 126, 140, 141, 143

I

independent power producers (IPPs), 15, 27, 98
industrial processes, 24, 39
insects, 5, 6, 8, 29-32, 38, 82, 107
International Boundary and Water Commission (IBWC), 130
International Joint Commission (IJC), 130

L

lakes, 3-7, 11, 14, 28, 37, 38, 46, 47, 50, 64, 74, 76, 82, 88, 90, 106, 112, 118, 121, 123, 126, 132, 133, 137, 141, 143, 152, 153
leaded gasoline, 31
Leaf Area Index (LAI), 52
levels of acidity, 5, 70
life-cycle assessment (LCA), 74
liquids- from-coal (LFC), 55
living things, vii, 1, 3
lower emissions rates, 25

M

megawatts (mw), 13, 23, 25
meteorological and chemical parameters, 42
microphysics, 42, 91
Monitoring Acid Rain Youth Program (MARYP), 95

N

National Acid Precipitation Assessment Program (NAPAP), 28, 29, 32, 33, 37, 54, 100, 102, 113, 115, 118, 129, 130, 152
National Aeronautics and Space Administration (NASA), 105
National Ambient Air Quality Standards (NAAQS), 131
National Association of Regulatory Utility Commissioners (NARUC), 115, 127
National Atmospheric Deposition Program, 2, 63
National Council of the Paper Industry for Air and Stream Improvement (NCASI), 33
National Emissions Data System (NEDS), 37
National Parks, 6, 14
National Stream Survey (NSS), 80, 91
National Surface Water Survey (NSWS), 4, 140
natural gas, xiii, xvii, 10, 11, 26, 31, 54, 56, 81, 109, 127, 134, 144, 154
natural stresses, 29, 32, 34
naturally occurring radioactive materials (NORM), 125
New Source Performance Standards (NSPS), 23, 25-28
nickel, ix, 31, 68, 132, 144
nitrate, xv, 9, 19, 49, 50, 58, 63, 69, 80, 81, 84, 85, 88, 91, 104, 121, 124, 142, 145, 149
nitric acid, 2, 87, 96, 99, 124, 141, 143
nitric oxide, 71, 78, 102, 125
nitrogen deposition, xiii, xviii, 6, 85
nitrogen dioxide, xii, xix, 24, 28, 127, 143, 145

nitrogen oxide(s) (NO$_x$), 1, 4, 9-14, 16-19, 21, 23, 26, 27, 29-32, 34, 35, 37, 40, 43, 47, 50, 52, 54, 55, 58, 59, 62, 66, 70, 77-79, 83, 84, 86, 87, 96, 100, 102, 103, 115, 117, 119, 125, 127, 128, 131, 140, 141, 147, 150, 152-154
Nutrient Cycling Model (NuCM), 89

O

octanol soluble carbon (OSC), xix, 94
oil, 10, 13, 31, 40, 50, 55, 56, 58, 72, 84, 89, 92, 109, 125, 126, 127, 134, 138, 140, 154
Opt-in Program, 15, 16
organic acids (OAs), 65
organic chemicals, 31
organic materials, 107
oxidants, 8, 32, 84

P

particulate organic carbon (POC), 81
persistent organic pollutants (POPs), 135
pH level(s), 4, 5, 38, 57, 60, 103, 117
Pine Barrens, 4, 113
planetary boundary layer (pbl), 47
pollutant emission controls, 29, 30
pollutants, x, xv, xix, 6-10, 29-34, 37, 40, 48, 55, 57, 58, 63, 65, 70, 73, 79, 84, 86-88, 90, 96, 99, 102, 103, 112, 132, 133, 135, 142, 143, 145, 147, 149
pollution control, 43, 76-78, 92, 96, 101, 110, 128, 135, 138, 145, 149, 151
powerplants, 25, 26
precipitation acidity, 69
process optimization unit (POU), 128
process-derived fuel (PDF), 55
public utility(ies) commissions (PUCs), 23, 89, 102, 138, 139

Q

quality assurance, 17, 133
quality control, 17, 133
quantitative index, 59

R

rain solution, 49, 98
rainband, 42
rainstorms, vii, 1
reactive hydrocarbons (RHC), 29, 31
reactive, xii, 3, 29, 31, 48, 50, 87, 112, 120
Reagan, President, 34

reduce emissions, 12, 16, 23, 53, 79, 102, 139
reduced growth, 30, 133
reduced inorganic sulfur (RIS), 113
reducing emissions, 13, 16, 58, 145
Regional Acid Deposition Model/Engineering Model (RADM/EM), 109
relative growth rate (RGR), 53, 117, 143
renewables portfolio standard (RPS), 45
Resource Conservation and Recovery Act (RCRA), 92, 126
resource management, 120, 130

S

Securities and Exchange Commission (SEC), 117
seedling exposure, 120
selective catalytic reduction (SCR), xviii, 102
self-consistent field (SCF), 122
sensitive forest soils, 3, 14
simulated acid fog (SAF), 85
simulated acid precipitation, 120
smog, 14, 21, 50, 57, 65, 84, 86, 99, 112, 113, 117, 118, 121, 140, 149, 154
snowmelt, xix, 5, 121, 132, 150
soil buffering capacity, 7
soil degradation, 6
soil sensitivity, 59
soluble reactive phosphorus (SRP), 112
Southern California Air Quality Study (SCAQS), 97
spatial and temporal variation, 109, 136
stomach acid simulants, 65
stratiform region, 42
stratospheric ozone, 38, 45, 50, 69, 82, 97, 108, 114
streams, xii, xv, xvi, xviii, 3, 4, 5, 7, 11, 14, 51, 59, 66, 69, 73, 80, 87, 90, 107, 113, 119, 140, 147, 152
streptomycin sulfate (SS), 150
stress factors, 30
sulfate production pathway, 109
sulfate, 8, 9, 19, 47, 49, 52, 55, 57-60, 62, 63, 65, 68, 69, 74, 80, 82, 84, 91, 95, 103, 104, 109, 111, 113, 117, 124-126, 130, 137, 138, 140, 141, 144, 145, 147, 149, 151
sulfur dioxide (SO$_2$), 1, 4, 6, 9, 10, 12-19, 22-32, 34, 35, 37, 39-41, 43, 44, 47, 50, 51, 53-58, 61-63, 65, 67, 68, 70, 73, 77, 79-81, 84, 86-88, 90, 92, 94, 95, 98, 100, 101, 104-107, 110, 112, 115-117, 120, 124, 126-131, 133, 134, 136-140, 142, 144-148, 152, 153
sulfur emissions, 47
sulfur gases, 43

sulfur oxides (SO_x), 31, 40, 55, 70, 75, 96, 119, 123, 129, 147
sulfur recovery, 47, 96
sulfuric acid, xiii, xiv, 2, 42, 54, 57, 60, 82, 87, 94, 102
sulfuric and nitric acids, 8, 31
sulphur, 76, 77, 90, 109, 132
surface waters, x, xvii, 4, 6, 20, 24, 28, 46, 58, 63, 66, 80, 93, 118, 141, 145

T

temperature-programmed desorption (TPD), 86, 87
thermally active marble (TAM), 116
titratable acidity, 59
tons per year (tpy), 27, 63, 109, 144
total nitrogen (TN), 68, 112
total organic carbon (TOC), 103
total phosphorus (TP), 112
toxic elements, 31
toxicity characteristic leaching procedure (TCLP), 96
tree foliage, 29, 32
trees, vii, xii, xiii, 1, 3, 6, 7, 14, 19, 29, 31, 32, 33, 35, 48, 60, 61, 63, 69, 85, 93, 103, 108, 116, 120, 121, 127, 131, 133, 136, 147, 150, 152
trichloroacetic acid (TCA), 150

troposphere, 47, 50, 56, 69, 119, 148, 149

U

urban areas, 86, 103, 144
US Department of Energy, 56, 64, 135

V

visibility reduction, 3, 9, 52, 152
Volume-weighted average (VWA), 95

W

waste management, 92, 112
water pollution, 73, 86, 97
water vapor, xvi, 8, 102, 127
watersheds, xi, xiii, xiv, xviii, 4, 58, 62, 69, 71, 93, 107, 132, 141, 145, 150
wet and dry deposition, 8, 19, 81, 125
wet deposition, 2, 20, 22, 32, 49, 66, 85, 150, 151
winter storms, 42

Y

yellowing and browning of foliage, 35